COPERNICUS

AND THE SCIENTIFIC REVOLUTION

By

EDWARD ROSEN
Distinguished Professor Emeritus
City University of New York, Graduate Center

AN ANVIL ORIGINAL

under the general editorship of

Louis L. Snyder

ROBERT E. KRIEGER PUBLISHING COMPANY
Malabar, Florida
1984

To Sandra

Original Edition 1984

Printed and Published by
ROBERT E. KRIEGER PUBLISHING COMPANY, INC.
KRIEGER DRIVE
MALABAR, FL 32950

Printed in the United States of America

Library of Congress Cataloging in Publication Data

Rosen, Edward, 1906-
Copernicus and the scientific revolution.

(Anvil Series)
"An Anvil original."
Bibliography: p.
Includes index.
Contents: Copernicus and the scientific revolution
--Readings.
1. Copernicus, Nicolaus, 1473-1543. 2. Astronomy--History--Sources. 3. Astronomers--Poland--Biography.
I. Title.
QB36.C8R65 1984 520′.92′4 [B] 83-9380
ISBN 0-89874-573-X (pbk.)

PREFACE

Under the auspices of the United Nations the entire civilized world in 1973 celebrated the five-hundredth anniversary of the birth of Nicholas Copernicus, the founder of modern astronomy. Specialists ransacked libraries and archives in a search for relevant sources that might have been previously overlooked. The older literature dealing with Copernicus and his times was reexamined with a view to supplementing and correcting it. As a result of these manifold efforts, an immense outpouring of books and articles in various languages enriched what was already known about the great astronomer.

Hence, the present is an auspicious occasion to review the older and newer contributions to our understanding of how Copernicus aided the advancement of mankind. The story of his life is unfolded, from his boyhood as an orphan to his recognition as a gifted thinker. The schools he attended; his professional career; his friends and foes; his hopes and fears; all these aspects of his biography are recounted and illustrated with appropriate documents, translated from the original languages, many for the first time. The outcome is a fresh view of Copernicus in relation to the Scientific Revolution, which since his time has grown to dominate and threaten the world in which we now live.

TABLE OF CONTENTS

Part I

COPERNICUS AND THE SCIENTIFIC REVOLUTION

CHAPTER ONE

THE EARLIEST ASTRONOMERS

Our earliest ancestors lived out in the open much more than most of us do today. To them the sky was like a familiar book. Sunrise and sunset, moonrise and moonset, the movements of the stars throughout the night, were closely linked with their everyday lives. They fed themselves mainly by killing and eating other animals. They stayed alive by avoiding being killed and devoured by other animals. To survive in this risky game of kill or be killed, they had to know when prey and predator prowled and slept.

The Phases of the Moon. The light that reached them at night varied much more than daylight. On certain nights, when the moon shone white and full, it was the brightest object in the sky. On other cloudless nights it could not be seen at all. In between these two extremes of complete visibility and total invisibility, it passed through a series of changes over and over again. After several nights when it was hidden, it first appeared in the western sky as a thin sliver in the form of a crescent. This grew wider night after night until after two weeks the whole round disk of the full moon came into view. Then it shrank gradually until it vanished from sight, when the whole sequence repeated itself. These phases of the moon constituted the synodic month. This was useful for keeping track of certain important biological cycles in nature.

The Seasons of the Year. Much shorter cycles could be measured conveniently by the time unit of a day - the interval between two successive risings or settings of the sun. Between two successive sunrises, daylight lasted longer than darkness during a certain period. Later on, during a corresponding period, darkness lasted longer than daylight. The transition occurrred on a day when the light lasted just as long as darkness. When this transition took place in the spring - the vernal equinox - the sun climbed to a certain height before beginning to turn down. The following day, at noon, when the sun reached its highest point for the day, it mounted a little higher than on the vernal equinox. This gradual daily ascent lasted three months until the day of longest light and shortest darkness - the summer solstice. Thereafter the sun's elevation at noon steadily decreased until, at the autumnal equinox, daylight equalled darkness once more. Then began the days of shorter and shorter daylight, and longer and longer darkness. After the winter solstice - the day of shortest daylight and longest darkness - the sun began again to rise higher at noon until the vernal equinox. Thus the four seasons of the year were completed. The breeding and migration times of land animals and fish were matched with the seasons. So were the food crops when our ancestors learned how to farm.

The Egyptian Solar Calendar. A very special kind of irrigational agriculture developed in Egypt, where the Nile River overflowed its banks with foreseeable regularity. The swift flood had to be controlled if human life and property were not to be swept away every year. River control required hard labor and careful planning in advance of the flood. Its start could be anticipated by the use of a calendar geared to the Nile's annual rise and fall.

Sirius, the brightest star in the Egyptian sky, could not be seen seventy days every year. During this period of invisibility from May to July it rose and set so close to sunrise and sunset that it was blotted out by the sunlight's immensely more powerful glare. This bit of astronomical information is known today, whereas the ancient Egyptians were not aware of it. They knew that on a certain morning each year Sirius began to be visible again above the eastern horizon just before the sun rose and made it fade out of sight. This is the day of Sirius' heliacal rising. The next day Sirius rose a little earlier. It separated itself from the sun a bit farther each day during its period of visibility from July to May, when it rose in the evening twilight. Then, after passing through its seventy-day period of invisibility, it re-emerged again on the day of its heliacal rising, quite close to the normal start of the Nile's inundation.

If the world had been made for human convenience, the year would consist of a whole number of uniform months, without any days left over. But the Egyptians found that Nature was not built that simply. They arbitrarily kept the number of days in every month at exactly thirty, disregarding the variations in the moon's actual travel time. Twelve of these arbitrary months would give them 360 days. To fill out the solar year, from one heliacal rising of Sirius to the next, after the end of the last month they tacked on five additional days not assigned to any month. Our present calendar is derived from the ancient Egyptian solar calendar. But our months are unequal, ranging from twenty-eight days in February to thirty-one days, say, in March. To avoid the inconvenience of five additional days not assigned to any month, we have had to give up the convenience of uniform months, with thirty days in each.

The Babylonian Lunar Calendar. In ancient Mesopotamia, where Iraq is today, the calendar was based on the moon. The interval from the first visibility of its crescent through all its other phases until the next visibility of the crescent varied. Sometimes this interval was twenty-nine days, sometimes thirty. Hence a calendar geared to the actual travel time of the moon would have been quite inconvenient. Regardless of how long any given month would actually be, the calendar adopted a standard month of thirty days. Twelve such months would fall short of a year. Whereas every year the Egyptians added five days not assigned to any month, the Babylonians preferred to insert a thirteenth month, when needed to keep the calendar in step with the seasons. This thirteenth month could not be intercalated every year, since 13 x 30 days = 390^d, much too long for the seasonal year.

The intercalation of a thirteenth month was therefore ordered by the authorities when required by Nature. But after centuries of spasmodic intercalations, the astronomers finally found a reasonably satisfactory equation. They assumed that in the long run the months would alternate between 29 and 30 days. Hence they could reckon with a schematic month of 29½ days. Then in 235 schematic months the days would number $6932\frac{1}{2}^d$, fairly close to 6935^d in 19 years of 365 days each.

This cycle of 19 years = 235 months was first instituted in 380 B.C. Of these 19 years, 12 received 12 months ($12^y \times 12^m = 144^m$), and the remaining 7 years received 13 months ($7^y \times 13^m = 91^m$; $144^m + 91^m = 235^m$).

The Islamic Calendar. The Islamic calendar today alternates 12 months of 29 and 30 days. This arrangement would equate the year with 354^d, a fraction less than the value accepted by the Muslims. Hence they use a 30-year cycle, in which 19 years are 354 days long, and the remaining 11 years are 355 days long. As a result 33 Islamic years = 32 Christian years. The Islamic calendar starts from Mohammed's flight (hijra) from Mecca in 622 A.D. Since that time 1361 Christian years have elapsed until the current year 1983, which matches 1403 Hijra.

The Eclipses of the Moon. Like Sirius, other stars rose somewhat earlier on successive nights. As they wheeled westward across the sky, the patterns they formed remained constant. Hence they were regarded as "fixed stars" - fixed in relation to one another. But the sun was thought to be slipping slowly eastward among them day by day in a journey that lasted a whole year. At the same time, every day of that year the sun traveled rapidly westward in the heavens in its diurnal journey.

Whenever the moon was eclipsed, either it was rising near sunset or setting near sunrise. At these times it was directly across the sky from the sun, and is said to be in opposition. But a lunar eclipse did not occur each time the moon was in opposition. It would undergo eclipse only when it was near the track followed by the sun in its annual eastward trek through the stars. This solar path was later named the "ecliptic", because the moon was eclipsed only when its opposition to the sun took place in the vicinity of the sun's yearly line of march. More often than not, the moon at opposition was not eclipsed. This was because it was then too far above or below the ecliptic. In those instances the moon's northern or southern latitude - its distance above or below the ecliptic- was too great for the effect to occur. However, when the moon approached one of its nodes, where its path crossed the sun's, it underwent an eclipse. After five or six months, there would be another eclipse.

The Babylonians did not regard the moon merely as a body in the sky. They worshipped it as a goddess, whose divine power could affect their lives.

> When at the moon's appearance its right horn is long and its left horn is short, the king's hand will conquer a land other than this. When the moon at its appearance is very large, an eclipse will take place. When the moon at its appearance is very bright, the crops of the land will prosper (R. Campbell Thompson, *The Reports of the Magicians and Astrologers of Nineveh and Babylon,* London, 1900, p. xxxviii, no. 30).

The Hebrews, however, recognized only one god. For them, the moon was a lesser light, just as the sun was a greater light. Hence, a Hebrew prophet scorned the "stargazers, those who reveal by the new moon what will happen to you" (Isaiah 47:13-14).

The Stars and the Planets. Among the many fixed stars moving across the night sky in unchanging patterns, a few bodies were detected wandering about. These wanderers - or planets, as they were later called - behaved differently from the stars. On the one hand, the stars moved steadily from east to west along a curved path night after night. On the other hand, while sharing in this motion, a planet shifted in relation to the stars. This shift was generally eastward, like the sun's annual motion and the moon's monthly motion. But whereas the sun and moon always moved eastward, the planet did not. It slowed down until it remained in the same place - its stationary point - several nights. During its station the planet continued to move westward in the nightly heavenly procession without changing its relation to the neighboring stars. But then the planet started shifting again, this time westward in relation to the stars. After a time it slowed down until it reached its second stationary point. Thereafter it resumed its eastward march.

This behavior of the planets attracted the attention of the Mesopotamian astronomers. They watched for a planet's reappearance after its period of invisibility. Then they marked its first station; its opposition to the sun, when it rose at sunset and was at its brightest; its second station; and its disappearance. Worshipping the planets as divinities, the Babylonian astronomers recorded these phases of their behavior.

> Two or three times during these days we have looked for Mars, but could not see it. If the king, my lord, should say "Is it an omen that it is invisible?" it is not (Thompson, p. xxxv, no. 21).

Yet these astronomers never asked themselves why these divinities disappeared and reappeared; why they stood still and moved again; and why they reversed themselves by moving in the retrograde direction.

Planetary Latitudes and Lunar Latitudes. The Babylonian astronomers recorded lunar eclipses accurately for centuries. By studying their records, they discovered the pattern in which eclipses duplicate themselves in a cycle of 223 synodic months, approximately equal to eighteen years. Yet these astronomers never found out - perhaps they never thought of the question - why the moon is eclipsed. They had learned from their numerous observations that the moon moved within a narrow range of 10° - 5° north and 5° south of the ecliptic. For them the moon's latitude was very important. For only when the moon came close to the ecliptic was it eclipsed. Their main interest in the moon was to be able to predict when it would be eclipsed. Their predictions could be right only if they accurately followed and foresaw the moon's motion in latitude.

The five planets then known also had motions in latitude. But the planetary latitudinal motions were not useful in predicting the invisibility of the planets, their reappearance, their first and second stations, and their retrogressions. Hence the Babylonian astronomers paid no attention to the planetary motions in latitude. They kept no record of them, by contrast with their careful and continuous information about the moon. Their interest was not to understand the inner workings of the cosmos. Their aim was to master those cosmic features which they believed controlled human destiny.

Greek Thinkers. The Greek philosopher Plato (427-c.348 B.C.) tried hard to figure out how various objects received their names. In particular, he remarked that the Greek word for the moon was connected with the idea of brightness. But as he said in his *Cratylus* (¶ 409), "it has recently been stated that the moon has its light from the sun." This magnificent discovery that the moon is really a dark body and shines only because it reflects the light it receives from the sun, was made by Anaxagoras, who died about the time when Plato was born. *(See Reading No. 1.)* This wonderful breakthrough came after centuries of uninterrupted observations of lunar eclipses by astronomers who never asked why the moon was eclipsed.

Anaxagoras, on the other hand, realized that the moon was

eclipsed when the sunlight heading toward it was intercepted by the earth. Hence, the edge of the earth's shadow falling on the moon during an eclipse gives a clue to the shape of the earth. The edge of that shadow is always curved. This is an observation that Anaxagoras missed. He still believed that the earth was flat and supported by the air, which it covered like a lid. But Aristotle (384-322 B.C.), a pupil of Plato, proved that the earth is round. He reasoned that loose heavy objects fall down - that is, toward the center - from every direction. As a result of all these vigorous pushes toward the center, the earth is pounded into the shape of a sphere. This physical argument was strengthened by two astronomical arguments. First, Aristotle pointed to the curved edge of the earth's shadow on the moon during an eclipse. Secondly, an observer traveling north or south notices corresponding shifts among the stars in the night sky. *(See Reading No. 2.)*

The darkness of the moon was used by Aristotle to prove that Mars is farther away from us than the moon is:

> We have seen the moon, when it was half-full, approach the planet Mars, blot it out behind the dark side of the moon, and let it emerge on the bright and shining side of the moon *(On the Heavens,* II, 12).

The Universe: Created or Eternal? In his treatise *On the Heavens* (I, 10) Aristotle argued against the doctrine of his teacher Plato that the heaven was created but is eternal. For Plato had written a dialog, the *Timaeus,* in which the main interlocutor was an astronomer of that name from Locri in Italy. In ¶ 280 B Plato had Timaeus ask:

> With regard to the entire heaven - or cosmos or whatever other more appropriate name it may receive, by which let us designate it...did it always exist, having no beginning of its coming into being, or did it come into existence by starting from some beginning?

Creation was the alternative accepted by Timaeus as more probable. In that case, what was created was unique. Only one world was created, not two or more or an infinite number. As Timaeus concluded, "There is and always will be only one." But he also added that this solitary world was alive, endowed with a World Soul, and had the shape of a sphere.

Differing with his teacher Plato as regards creation and the World Soul, Aristotle wrote in the *Heavens* (II, 1):

> The universe as a whole neither came into being nor can it be destroyed....On the contrary, it is unique and everlasting. It has neither a beginning nor an end of its entire existence, and it also possesses and contains within itself time without end.

Both Plato and Aristotle lived and died long before Christianity originated. As it developed, it felt the need to strengthen its message by absorbing the teachings of the outstanding Greek philosophers. Plato was particularly suitable for this purpose. His account of the creation of the universe could be blended somehow with both creation stories in the Hebrew Bible, which was also adopted by Christianity under the name "Old Testament." Unlike the gods worshipped by the Greeks, Plato's creator did not have an individual name. He was simply called the Craftsman, or Demiourge in Greek. Later on, when Aristotle's philosophical system seemed more appropriate to Christian theologians, they had to find some way around his belief that the universe never had a beginning, since his eternal universe was incompatible with the Biblical account of creation. In some parts of the United States a similar struggle is being waged today between those who accept the Biblical account of the creation of living things, and those who believe in the origin of new species through an evolutionary process of natural selection.

Why do the Stars Shine? According to Plato's creation story in the *Timaeus* (¶ 39b),

> In the second orbit from the earth, the god kindled a light, which we now call the sun, especially to illuminate the heaven throughout.

Plato mentioned the creation of no light other than the sun. Hence, some of his followers generalized from the case of the moon. They believed that all the other heavenly bodies - the stars and the planets - were like the moon, dark in themselves, and lit up by the sun, the unique source of light in the entire universe.

Aristotle disagreed. In the *Heavens* (II,7) he wrote that "The stars' heat and light are generated by the friction between their movement and the air." He applied the same reasoning to the sun, in his *Meteorology* (I, 3; 341a):

> With regard to the origin of the heat furnished by the sun...it may now be explained why it is generated, such bodies not being [hot] by nature. We see that motion can thin the air out and set it on fire, so that moving objects are often observed even to melt. For the production of warmth and heat, the sun's motion alone is enough for this result. For, the motion must be swift and not far away. The motion of the stars is swift, but far away. The moon's motion is near, but slow. The sun's motion possesses enough of both speed and proximity.

Some Platonists thought that the stars were not self-luminous. Aristotle stated that the heavenly bodies became visible only because of the friction generated by their motion. It was not until the telescope was invented and turned toward the heavens that the real difference between the stars and the planets was discovered: the stars shine by their own light, whereas the planets are like the moon and shine only by reflected light. But just before the invention of the telescope, a brilliant Italian thinker realized the true nature of the sun: it is a star, and shines by its own light, like the other stars. We on earth feel its heat because we are much closer to the sun than to any other star.

Two Babylonian Presentations of Planetary Motion. Besides sharing in the nightly westward procession across the sky, the planets differentiate themselves from the neighboring stars by undergoing stations and retrogressions. This behavior of the planets was recorded by the Babylonian astronomers. In their wedge-shaped or cuneiform script, they impressed on dampened clay tablets long columns of numbers, which carried no explanatory headings. In a remarkable burst of scholarly ingenuity during the past century these astronomical tablets were deciphered, and their hidden meaning was recovered.

For instance, a sequence of years is listed by a tablet in its first column, with the month and day in the third column. The fifth and last column gives celestial positions on the day and month indicated in the third column, within the year recorded in the first column. The differences between successive entries in the fifth column are noted in the fourth column. There, the first seven differences show a decrease of 12° ; the next fourteen, an increase of 12° ; the next fifteen, a decrease of 12° ; and the last six, an increase of 12°.

An annual motion of 12° , after thirty years ($12^{\circ} \times 30^{y} = 360^{\circ}$), restores a planet to its original position. In this tablet the planet is recorded at 11° in the sign of the Goat, and thirty years later at 18° in the same sign. This pattern of about 360° in 30 years fits the planet Saturn. At the beginning of the calendar year Saturn was in the sign of the Balance, directly opposite the sun in the sign of the Ram. Being six signs away from the sun, Saturn rose as the sun set. Accordingly, despite the absence of labels, oppositions of Saturn are recorded by this tablet. It has given up its secret to the skilled specialists in cuneiform astronomy.[1]

As Saturn performs its 30-year cycle, its speed varies. Instead of trying to track its actual varying speed, the Babylonian astronomers assigned it alternating decreases and increases during successive clusters of years: 12° less, 12° more. The result may be diagrammed as a linear zigzag pattern:

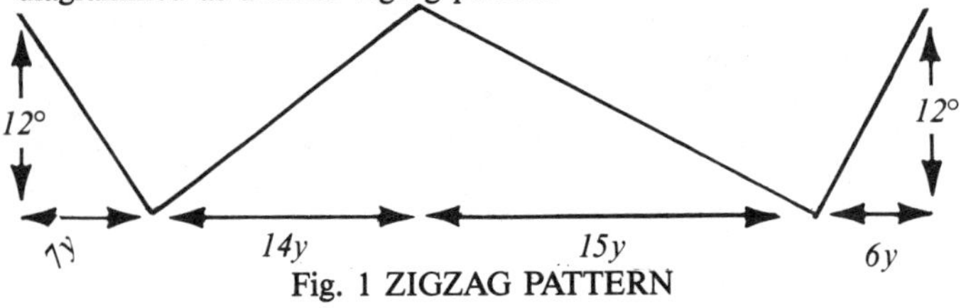

Fig. 1 ZIGZAG PATTERN

Alternatively, the speed could be held constant at a given level, then discontinuously dropped to a lower level, and later jumped back up again to the original level. This computational arrangement assumes what is called a step pattern:

Fig. 2 STEP PATTERN

Both Babylonian patterns, the step and the zigzag, were exclusively arithmetical in nature. No three-dimensional models were visualized. The heavens were not conceived as the upper part of a vault rotating around a pivotal point. The concept of a spherical heaven, and circular patterns of motion for the stars and planets were introduced by the Greeks.

Planetary Loops and Circular Rotations. As a planet moves eastward against the background of the stars, its speed varies until it reaches its first stationary point. There it reverses its direction and moves westward until it reaches its second stationary point, where it reverses itself again and resumes its direct march eastward. Between the two stationary points it executes a loop.

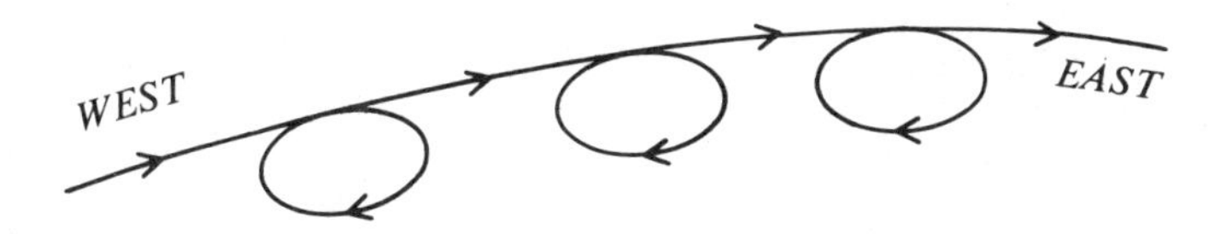

Fig. 3 PLANETARY LOOPS

These loops and changes in direction and speed might seem suitable for a dancing girl in the countryside. But they were hardly deemed worthy of a divinity, for a planet was so regarded by the Greeks. In particular, Plato was intensely religious. For those people who were by nature absolutely honest, but did not believe in the gods, he prescribed the death penalty.[2]

Plato himself made no contribution to mathematics. But he had a profound appreciation of its importance. He regarded the circle as the most perfect of all two-dimensional geometric figures. It had neither a beginning nor an end. If a god-planet moved in a circle, it would never change direction. It would have no motive to speed up, slow down or stop. It could keep going forever, as Plato thought the universe would. But an actual planet, as observed, does not pursue a circular path. Perhaps an inventive mathematician, Plato suggested, could devise a combination of circular rotations that would produce the planetary loops.

Plato and Simplicius. Plato did not make this suggestion in any of his numerous dialogs. These were studied with the utmost care by his followers in the Academy he founded. But in 529 A.D. the Christian emperor Justinian forbade pagans to teach. Hence Simplicius, a Neoplatonist, as Plato's followers were later called, had to give up his profession. But he continued to write his commentaries on the philosophical classics. In his *Commentary on Aristotle's Treatise on the Heavens* Simplicius recalled Plato's suggestion about combining uniform circular rotations to produce the nonuniform and noncircular planetary loops.

Although this suggestion is not present in any of Plato's dialogs, there is no good reason to doubt the word of Simplicius. His quotations from earlier works that have survived can be checked today and they are uniformly accurate. In fact, a quotation in Simplicius is used now to show what reading he had in his copy of a book, as against a variant reading elsewhere. Printing had not yet been invented. Copies of a printed book do not differ. But before the invention of printing, when books were copied and recopied by hand, variant readings crept in for all sorts of reasons. In his extensive library Simplicius had books which have since disappeared. He tells us where he found Plato's suggestion in works no longer available. *(See Reading No. 3).*

The Concentric Spheres of Eudoxus. Plato's suggestion was first acted on by his pupil and associate, the brilliant mathematician Eudoxus. He imagined the planet placed at a point on the equator of a sphere revolving at constant speed. But the poles of this carrying sphere were attached to another sphere revolving about different poles in the opposite direction. The combined effect of these two spheres carried the planet through a figure-of-eight pattern lying on its side. These two inner spheres were at the same time transported eastward by a third sphere at the speed of the planet's orbital motion. Finally, the fourth and last sphere was linked to the sphere of the stars performing the daily rotation of the heavens from east to west.

For each of the five planets then known, four spheres were needed, making a sub-total of twenty. For the sun and moon, three spheres were enough. The sphere of the stars brought the total up to twenty-seven. All of Eudoxus' twenty-seven spheres had the same center - the center of the earth at the center of the universe. Hence these spheres came to be known as concentric, the Latin equivalent of the Greek "homocentric."

Callippus' Improvement of Eudoxus' Concentric Spheres. Eudoxus' twenty-seven concentric spheres - the earliest attempt to clear up the puzzle of the planets - failed to solve the problem. Hence, "Callippus of Cyzicus...went to Athens, stayed with Aristotle, and together with Aristotle corrected and completed Eudoxus' discoveries."[3] Aristotle himself took no credit for these improvements in Eudoxus. In his *Metaphysics* (XII,8) Aristotle says:

> Callippus...assigned the same number of spheres to Jupiter and Saturn as Eudoxus did [that is, four]. But Callippus believed that two more spheres should be added to the sun and to the moon, if one wants to account for the phenomena, and one more to each of the remaining planets,

making four spheres for the sun, moon, Mars, Venus, and Mercury. Thus, Callippus made the total number of spheres thirty-four: one for the stars, four each for Jupiter and Saturn, and five each for the sun, moon, Mars, Venus, and Mercury.

Just as Aristotle claimed no credit for these additional spheres, so he offered no explanation why Callippus thought they were necessary. As Simplicius says,

> No writing of Callippus is known giving the reason for these added spheres, nor did Aristotle give it. But Eudemus referred concisely to the phenomena on account of which Callippus thought these spheres should be added. According to Eudemus, Callippus said that...the intervals between the solstices and equinoxes vary....The three spheres [of Eudoxus] are not enough to account for the phenomena of each [of these bodies, the sun and the moon], on account of the observed irregularity in their motions. As for the single sphere which Callippus added to each of the three planets, Mars, Venus, and Mercury, Eudemus explained compactly and clearly why it was added.[4]

Unfortunately, Simplicius did not repeat Eudemus' explanation. Hence, modern historians of astronomy must conjecture exactly how and why Callippus' additional sphere improved Eudoxus' theory for Mars, Venus, and Mercury.

Aristotle's Mechanization of the Concentric Spheres. For each planet both Eudoxus and Callippus devised a set of concentric spheres that were not connected in any way with any other planet's concentric spheres. As astronomers, Eudoxus and Callippus tried to explain the path of each planet separately by purely geometrical constructions. But Aristotle the philosopher was a system-builder. He was not satisfied to have any of his planets go its own way. For him, there was only one universe, and all its parts must fit together and work together. Hence the spheres controlling the motion of any planet would affect the planet just below it, and introduce distortions in its motion.

In order to avoid such disturbances, Aristotle thought that the lower planet must be provided with spheres that would counteract the effect on it of the upper planet's spheres. The upper planet was carried on the innermost of its spheres. Between this innermost sphere of the upper planet and the outermost sphere of the lower planet, Aristotle interposed his counteracting spheres. Jupiter, for example, must have three counteracting spheres to cancel out Saturn's figure-of-eight and orbital motion. But there was no need to eliminate Saturn's fourth and innermost sphere, participating in the heavens' daily rotation. Hence, in addition to its own four spheres, Jupiter acquired from Aristotle three more spheres to counteract Saturn's effect on it.

The same reasoning was applied all the way down the line, where four counteracting spheres were added for Mars, Venus, Mercury, and the sun. Hence, to Callippus' thirty-three spheres carrying the planets, Aristotle added three counteracting spheres for Saturn and Jupiter, plus four for Mars, Venus, Mercury, and the sun. Thus, Aristotle emerged with a total of fifty-five spheres for the planets. *(See Reading No. 4).* He omitted counteracting spheres for the moon and included them for Saturn, for reasons which he did not explain. Because of Aristotle's enormous prestige, his fifty-five spheres exerted a powerful influence for centuries.

The Fatal Defect of the Concentric Spheres. Whether in the form devised by Eudoxus, improved by Callippus, or mechanized by Aristotle, the system of concentric spheres always placed a planet at the same distance from the earth. For in this concentric system the earth was the universe's center, which was also the center of all the concentric spheres. Hence, a planet was always at exactly the same distance from the earth, in whatever direction the planet was seen. Being divine, it was not subject to change. In particular, it could not become bigger or smaller. Yet the moon sometimes looks bigger and brighter, sometimes smaller and fainter. So does Venus. So does Mars. Since these bodies never change their size, yet look bigger and smaller, their distance from the earth must vary. But their distance from the earth in the system of concentric spheres cannot vary. This is the fatal defect in the system of concentric spheres. *(See Reading No. 5.)*

Notes

1. This tablet is transcribed as No. 702 in *Astronomical Cuneiform Texts,* ed. O. Neugebauer (London, 1955), III, 207; see also Neugebauer's *History of Ancient Mathematical Astronomy* (New York, 1975), II, 381.

2. Plato, *Laws,* X, 908-909.

3. *Simplicii in Aristotelis De caelo commentarii,* p. 493/5-8.

4. *Ibid.*, p. 497/15-24.

CHAPTER 2

GREEK ASTRONOMY AT ITS HEIGHT

Babylonian astronomers were scrupulously attentive to the particular phenomena that interested them. What happened between those critical points concerned them less. Instead of making careful observations all along the line, they watched for crucial turns. About the intervals in between, they made easy assumptions. Thus, they strained themselves to find out how long a year is. But they took for granted that each season had exactly the same length.

Under their influence, after the Hebrews had been exiled to Babylonia, in the third century after Christ, Rabbi Samuel, who was famous as an astronomer, stated: "A season of the year has no more than 91 days and 7½ hours." The same interval, $91^{d}\ 7½^{h}$, defined the time between an equinox and a solstice, and a solstice and an equinox, according to Rabbi Hanina.[5]

This conclusion was not reached by observation. It was simply the result of dividing $365¼^{d}$ as the length of the year by 4, the number of the seasons.

Papyrus Paris No. 1. The ancient Egyptians made a kind of writing paper from a local plant. Some of these Egyptian papyri have survived, more or less intact, to our day. One of them, now labeled Papyrus No. 1 in the Louvre in Paris, was written early in the second century B.C., after the Greeks had conquered Egypt. This particular papyrus, written in Greek, looks like a brief elementary school notebook about astronomy. Near the end, it gives a whole number of days as the interval:

	According to	
	Euctemon	Callippus
From the summer solstice to the autumnal equinox	90	92
From the autumnal equinox to the winter solstice	90	89
From the winter solstice to the vernal equinox	92	90

The fourth and last line is missing, but can be reconstructed:

From the vernal equinox to the summer solstice	93	94

Both Euctemon (about 430 B.C.) and Callippus a century later recognized that the four seasons were unequal in length. In this respect they were right, even though their numerical results differ from modern computations.

The Length of the Year and the Seasons. In Papyrus Paris No. 1 the year is equated with a whole number of days, 365. But an additional fraction of a day is required. To discover the amount of this fraction, a later Greek astronomer compared his own observation of the summer solstice with the oldest corresponding observation he could find. This had been made on 27 June 432 B.C. by the school of Euctemon. This comparison led to the conclusion that the year contained, in addition to 365d, $\frac{1}{4}^{d}$ - $1/300^{d}$. This fractional total consisted of four seasons of unequal length. Thus, the interval from the vernal equinox to the summer solstice was $94\frac{1}{2}^{d}$, and from the summer solstice to the autumnal equinox $92\frac{1}{2}^{d}$, with the autumn and winter $88\ 1/8^{d}$, $90\ 1/8^{d}$, making the total $365\ \frac{1}{4}^{d}$.

Eccentric Motion. According to the Greek thinkers, the sun revolved around the earth in a circle at uniform speed. If the earth were at the center of the sun's circle, the four seasons would have to be equally long. But they are not. Hence, the earth could not be at the center of the sun's circle. In other words, the sun moved along a circle, the center of which was outside the earth. Such a circle was called eccentric.

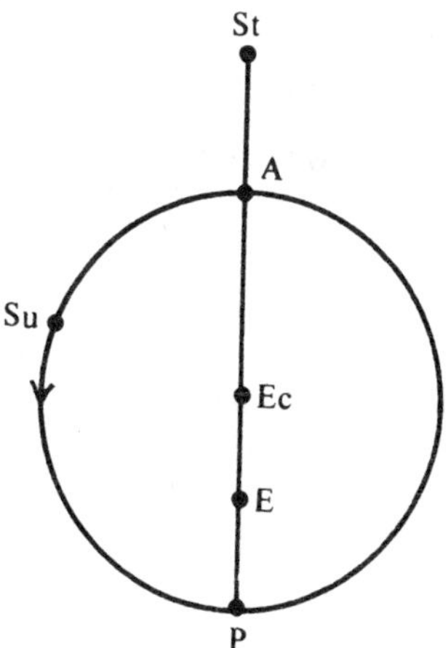

Fig. 4 ECCENTRIC CIRCLE

In Fig. 4 the sun Su revolves around an eccentric circle centered at Ec. The earth E is stationary at the center of the universe. The distance from Ec to E is the eccentricity. A line drawn from E through Ec may be extended to the firmament, which it will meet at or near a star St. This line intersects the eccentric circle at A. When the sun in its motion around the eccentric circle is at A, it is at its farthest distance, or apogee, from the earth. Diametrically opposite A is P, where the sun is in its perigee, or closest approach to the earth. As the sun moves around the eccentric circle, its motion is uniform with respect to Ec, but nonuniform with respect to E. Its uniform motion around Ec looks nonuniform to observers on E, the earth, because they are watching from a point outside the center of the eccentric circle. Hence, when the eccentricity is correctly chosen, the sun's motion on the eccentric can exactly match the unequal length of the seasons.

Epicyclic Motion. Place a planet on the circumference of a circle. The center of this circle is located on the circumference of a second circle. As this carrying circle, or deferent, turns, the center of the surmounting circle, or epicycle, turns with it. At the same time, the epicycle itself turns in the opposite direction. An earthling on E, the center of the deferent, observes the planet. While it moves from P_1 to P_2, it pursues its direct motion. But from P_2 to P_1 it pursues a retrograde motion. Between P_2 and P_1 it passes through its perigee; between P_1 and P_2 it passes through its apogee.

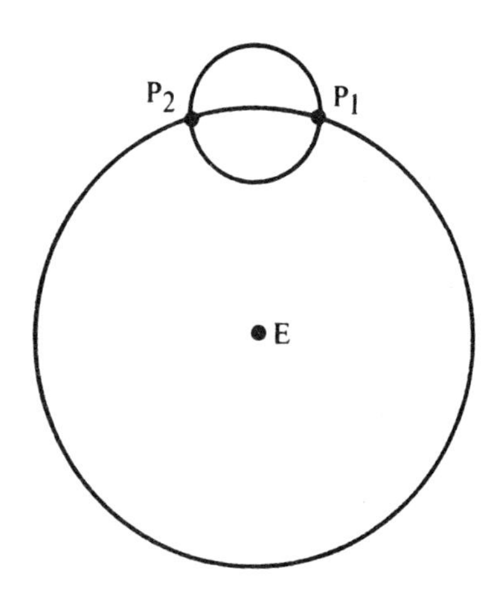

Fig. 5 EPICYCLE AND DEFERENT

Thus, the combination of deferent-plus-epicycle could explain the variation in the brightness of the planets as well as their loops.

The Equivalence of Eccentric and Epicyclic Motion. The geometrical and physical equivalence of eccentric and epicyclic motion was an important discovery of the Greek astronomers. Put Fig. 4's eccentricity Ec - E equal to Fig. 5's radius of the epicycle. In Fig. 6 the dotted circles are the deferent and epicycle, while the solid circle is the eccentric. Start the planet's motion at P, where the eccentric intersects the epicycle. Rotate the eccentric and the deferent in one direction, and the epicycle in the opposite direction, at the same speed. Then the planet will return to P at the same time, whether it is carried by the eccentric or by the deferent-plus-epicycle. In either arrangement the planet will always reach the same point at the same time.

The Greek astronomers could therefore choose either the eccentric model or the deferent-plus-epicycle model in any given case. Their choice depended on the suitability of the model to the observational data available to them. Under these circumstances they recognized that neither model was physically real. Each was mathematically useful. In the same way, a modern seafarer does not expect his ship to hit a bump when it crosses the equator. Nevertheless, the equator is a useful concept in geography.

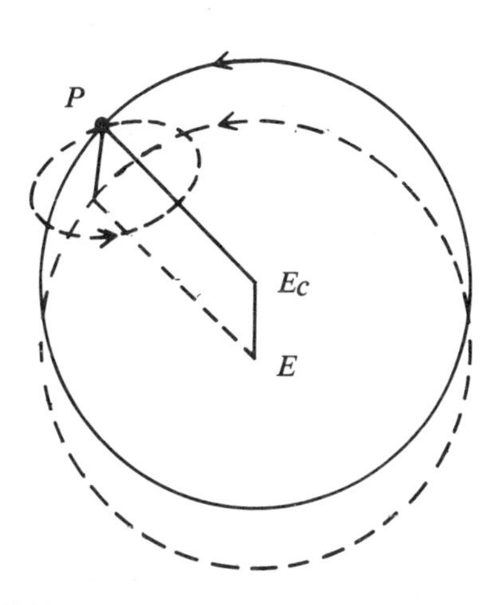

Fig. 6 EQUIVALENCE OF ECCENTRIC AND EPICYCLIC MOTION

The Equant. The Greek astronomers showed great skill in manipulating the eccentric, deferent, and epicycle, separately or in combination, to match the observed motions of the heavenly bodies. But Ptolemy, who was active in the second century after Christ as the greatest of the Greek astronomers, introduced a refinement in planetary theory. In Fig. 7's eccentric circle he drew a diameter through E, the center of the earth, and Ec, the center of the eccentric. At the distance of this eccentricity (E - Ec) he placed a point Eq, so that E - Ec = Ec - Eq. A planet moving on the circumference of the eccentric was always at the same distance from the eccentric's center, Ec. But, as measured from Ec, the planet's speed varied. The speed, however, was always uniform with respect to Eq, later called the equalizing point or equant. Ptolemy's planet revolved with uniform speed around the equant, from which the planet's distance varied. Its distance was invariant, however, from the eccentric's center, about which its speed varied. By having a planet move at a constant distance from one point but with uniform speed around a different point, Ptolemy broke sharply away from the previous requirement that circular motion must be uniform around its own center. By adding the equant to the eccentric, deferent, and epicycle, Ptolemy was able to surpass all his predecessors in bringing his planetary theory into closer agreement with his observational data. The Ptolemaic system reigned supreme in astronomy more than a thousand years.

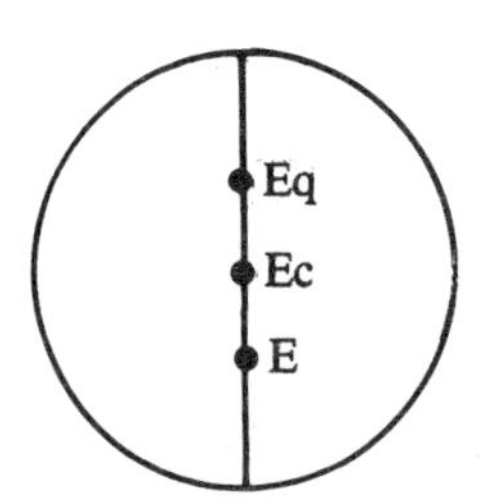

Fig. 7 EQUANT

The Earth, Movable or Immovable? Book 2, Chapter 13, of Aristotle's *Heavens* begins by saying: "It remains to talk about the earth, where it happens to be, and whether it belongs to things at rest or in motion." Aristotle then raises the question why a

> small portion of earth, if lifted up and released, moves without preferring to remain still; the bigger it is, the faster it moves; yet if anybody raised the entire earth and let it go, it would not move. Actually, heavy as it is, it is at rest.

Aristotle's earth is at rest in the center of the universe.

> Not only is it seen resting at the center, but also moving toward the center. For wherever a part of it moves, the whole must move there too. Whither it moves by nature, there it also remains by nature.

Then in Book 2, Chapter 14, Aristotle adds:

> The motion of the whole earth and of its parts is by nature toward the center of the universe. For that reason the earth happens to be lying now at that center....It happens that the earth and the universe have the same center....The earth does not move nor does it lie outside that center.

Heraclides of Pontus. From Pontus, a district on the southern shore of the Black Sea, Heraclides went to Athens, where he studied with both Plato and Aristotle. Heraclides agreed that the earth was centrally located. But the earth could move wihout leaving its central location. Let it remain in its place at the center of the universe. But let it rotate once a day about its own axis from west to east. At the same time take this daily rotation away from the heavens. Then the daily rising and setting of the sun, moon, stars, and planets would look the same to an observer on the earth as they would if he and the earth were motionless, and the heavens moved daily from east to west.

Heraclides wrote many books, but not a single one has come down to us. Nevertheless, his belief in the daily rotation of the earth was recorded in a work called *The Opinions of the Philosophers.* These snippets were assembled by an unknown compiler. Like many another faker of ancient literary frauds, he falsely ascribed his collection to a well-known writer. The pretended author chosen for the *Opinions of the Philosophers* was Plutarch, the immensely popular narrator of the *Parallel Lives* of Greeks and Romans. Pseudo-Plutarch says in Book 3, Chapter 13, of the *Opinions* that

> Heraclides of Pontus...makes the earth move, not in a progressive motion, but like a wheel in a rotation from west to east about its own center.

Ecphantus the Pythagorean. Coupled with Heraclides in this quotation from Pseudo-Plutarch is Ecphantus the Pythagorean. Pythagoras founded a philosophical and religious fraternity, sworn to secrecy. Hence Ecphantus did not commit his views to writing. After visiting the Pythagoreans in Sicily, Plato made Timaeus the Pythagorean the principal speaker in his dialog of that name. In the manner of his teacher, Plato, Heraclides also wrote dialogs, now lost. One of them was entitled "On the Pythagoreans." There Heraclides may have used Ecphantus as a speaker to expound the daily rotation of the earth. As a result the speaker Ecphantus in Heraclides' "On the Pythagoreans" was later coupled with Heraclides as believing in the earth's rotation.

Archimedes' *Sand-Reckoner.* The Greeks used the letters of their alphabet to denote the numbers from 1 to 999. For thousands, they put a short dash in front of the appropriate letter. For tens of thousands they used the abbreviation M for "myriad." But how could they write a number expressing all the grains of sand in the universe?

How big is the universe or cosmos? This question was answered in two different ways by Archimedes, the greatest mathematician in antiquity, who was killed by an enemy soldier when the Romans captured Syracuse in Sicily in 212 B.C. In his *Sand-reckoner* Archimedes explained that most astronomers then regarded the universe as a sphere. Its center was occupied by the center of the earth. A line drawn from the center of the earth to the center of the sun was the radius of the universe, as envisaged by most Greek astronomers in the time of Archimedes.

Aristarchus of Samos. Archimedes' second answer to the question about the size of the universe was taken from Aristarchus of Samos, who was about twenty-five years older than Archimedes. Aristarchus published certain hypotheses making the universe many times bigger than the universe as conceived by his contemporaries. Aristarchus' hypotheses have survived only because they were summarized in Archimedes' *Sand-reckoner*:

> The fixed stars and the sun remain stationary. On the other hand, the earth revolves around the sun along the circumference of a circle lying amid the course [of the planets]. The sphere of the fixed stars has the same center as the sun.

Hence Aristarchus put the sun in the center of the universe. He is therefore the originator of the heliocentric astronomy. His sun is stationary. He is therefore the originator of the heliostatic astronomy. His earth revolves around the stationary sun. By this geokinetic revolution he could have explained the planetary loops. His stars are stationary. Therefore, he attributed the daily rotation to the earth as a second geokinetic motion. This is not specified by Archimedes, whose attention in the *Sand-reckoner* was focused on the size of the universe.

Plutarch, however, wrote a dialog about *The Face on the Moon.* One of the speakers in it recalls (¶ 923) that Aristarchus

> assumed that the heaven is stationary, while the earth revolves along an inclined circle and at the same time rotates around its own axis.

The testimony of Archimedes and Plutarch makes it absolutely clear that the heliocentric, geokinetic astronomy was proposed by Aristarchus of Samos. No hint of his views was found in Babylonia, until about a century after him Seleucus, from Seleuceia on the Tigris River, "also affirmed the motion of the earth," according to

Pseudo-Plutarch, III, 17.

The Rejection of Aristarchus. Apart from Seleucus, Aristarchus attracted no support from the Greeks. But he did find opponents. A contemporary philosopher, the head of the Stoics, filled with religious fervor, "thought it was the duty of the Greeks to indict Aristarchus of Samos on the charge of impiety for putting the earth in motion."[6] A later Platonic philosopher and astrologer said "that we must believe the earth...remains stationary...For the planets, together with the entire heaven which contains them, are in motion. He rejects with abhorrence as opposed to the hypotheses of divination those who made moving bodies still, and set in motion those bodies which are motionless by nature and position."[7]

Ptolemy's Opposition to Geokineticism. More influential than the resistance of philosophers to Aristarchus was Ptolemy's antagonism toward his geokinetic views. The philosopher who was Aristarchus' contemporary deemed it advisable to mention his name while charging him with impiety. The later philosopher who abhorred his views did not single him out, because his hypotheses were no longer available as such. This was also Ptolemy's strategy.

Without mentioning Aristarchus by name, in his *Mathematical Syntaxis* Ptolemy devoted a whole chapter (Book 1, Chapter 7) to contending "that the earth performs no progressive motion." By "progressive motion" Ptolemy meant travel from one place to another. He insisted that the earth occupied the center of the universe without ever leaving it. But what about the earth rotating daily around its own axis while still remaining in the center of the universe? Ptolemy called this a "simpler arrangement." He conceded that "as regards the heavenly phenomena, perhaps nothing would prevent the situation from being in agreement with" it.

The Air Around the Earth. What happens on the earth and near it was the basis of Ptolemy's objection to the earth's daily rotation. A heavy object dropped from a height falls toward the earth. In the air a heavier object falls faster. As a whole, the earth is heavier than any of its parts. Therefore, it must fall faster than any of its parts, however heavy. The earth "would swiftly drop out of the heavens altogether." Living creatures and partly dense objects would be left hanging in the air.

If the earth rotated in place, any point on its surface would have to whirl through a vast distance eastward every twenty-four hours. Any light object on earth could not keep up with such speed, and would be seen traveling westward. All clouds would drift in the same direction. Any object thrown vertically upward would not come down to the place from which it was launched. For, that place would have moved eastward during the object's ascent and descent.

A geokineticist could respond that the air surrounding the earth rotated with it in the same direction and with the same speed.

Then all the objects in the air would be left behind, and be seen always moving westward. But we do not observe any such unidirectional westward procession in the air. Alternatively, the geokineticist might reply that airborne objects shared in the motion of the air. Then they would all move with the speed of the air. Their position above the earth would never change. They would not accelerate, slow down, or shift direction. Again, such a permanently unchanging procession is not what we see in the air above us. Such was Ptolemy's reasoning as he enshrined the stationary earth. *(See Reading No. 6.)*

China. Chinese astronomers were not aware that the earth is round. They were equally unaware of its motion. There was one exception, in the first century B.C. If the earth moves, why do not we earth-dwellers feel that motion? This question was not answered directly, but an imperfect analogy was proposed:

> The earth is constantly in motion, never stopping, but men do not know it: they are like people sitting in a huge boat with the windows closed; the boat moves but those inside feel nothing.[8]

We earth-dwellers do not feel the motion of our earth. But we are not shut in by closed windows. Unlike the travelers in the Chinese boat, we can see outside our spacecraft. Yet it is not easy to understand what we see.

India: Aryabhata. No hint of geokineticism is found in native Babylonian astronomy. The same may be said of Egyptian astronomy. On our side of the world, among the Maya before Columbus, a skillful and intricate calendar astronomy developed which never referred to the earth as a moving body. The Sanskrit literature of ancient India was equally silent about geokineticism until after Alexander the Great, king of Macedonia, overthrew the Persian Empire and penetrated what is now Pakistan.

In A.D. 499 Aryabhata, who was then twenty-three years old, wrote a poem entitled *Aryabhatiya,* which is one of the earliest Sanskrit treatises on astronomy still preserved. In Chapter 4, verse 9, Aryabhata said:

> As a man in a boat going forward sees a stationary object moving backward, just so at Lanka [the earth's equator] a man sees the stationary asterisms [stars] moving backward [westward] in a straight line.[9]

Aryabhata never thought of the earth's annual revolution. Among the Indian astronomers, he was the principal proponent of the earth's daily rotation. For this reason he was severely criticized even by some who followed his other ideas. Hence a certain suspicion hangs over Aryabhata's Chapter 4, verse 10. This comes right after verse 9, which was just quoted with its reference to the stars as stationary. Yet verse 10 explains that the stars rise and set because their circle "constantly moves straight westward." Could some misguided supporter of Aryabhata have changed verse 10 to bring the master into conformity with the mainstream of Indian tradition? In Chapter 1, verse 4, Aryabhata explained how far the earth moves in a minute. But this text was later altered in order to eliminate the earth's motion. Prior to the alteration, however, Aryabhata's authentic text referring to the earth's motion was quoted in 628 by Brahmagupta.

Brahmagupta. Brahmagupta's main treatise was studied thoroughly by al-Biruni (973 - c. 1050), the best informed of the medieval Muslim writers. When his own country was overrun, he was forced to live in Afghanistan. His new ruler's further conquests in India, however, let him travel in that subcontinent. *India* was his longest and greatest work. Having learned Sanskrit, he translated from that language into Arabic. For, he preferred Arabic as the vehicle of science to his own Persian mother tongue. In Chapter 26 of his *India,* al-Biruni reported:

> As regards the resting of the earth, one of the elementary problems of astronomy, which offers many and great difficulties, this, too, is a dogma with the Hindu astronomers. Brahmagupta says..."Some people maintain that the first motion [from east to west] does not lie in the meridian, but belongs to the earth."

In opposition to these geokineticists, al-Biruni quoted a sixth-century Sanskrit astronomer who refutes them by saying: "If that were the case, a bird would not return to its nest as soon as it had flown away from it towards the west."[10]

Al-Biruni concurred: "And, in fact it is precisely as [this sixth-century astronomer] says." But al-Biruni continued this discussion by quoting again from Brahmagupta, who says in another place of the same book: "The followers of Aryabhata maintain that the earth is moving and heaven resting. People have tried to refute them by saying that, if such were the case, stones and trees would fall from the earth." But Brahmagupta does not agree with them, and says that that would not necessarily follow from their theory, apparently because he thought that all heavy things are attracted towards the center of the earth.

Was Brahmagupta's implication correctly understood by al-Biruni? Was al-Biruni's Arabic translated accurately into English? If so, Brahmagupta's implication effectively answered Ptolemy's main objection to geokineticism: heavy objects do not fall off the earth because they are attracted to it.

Brahmagupta's thinking did not convince al-Biruni. He ended Chapter 26 of his *India* by concluding that for reasons outside of astronomy the earth cannot move. The translator of al-Biruni,[11] however, cautioned (I,lxv) that Arabic "sometimes exhibits sentences perfectly clear as to the meaning of every single word and the syntactic construction, and nevertheless admitting of entirely different interpretations." In the translator's interpretation, al-Biruni rejected the earth's rotation for physical, not astronomical, reasons.

No such definite stand was taken by Seneca, the Roman philosopher who tutored Nero. After his pupil became the emperor of Rome, he ordered his former tutor to commit suicide in April A.D. 65. Not long before that time Seneca wrote in his *Natural Questions* (7,2) that we ought to

> know whether the earth stands still while the heaven rotates, or the earth turns while the heaven stands still. For there were those who said that it is we who are carried around by nature without our knowing it; and that risings and settings do not happen because the sky moves, but we ourselves rise and set. This is a subject worth investigating in order that we may know what our situation is: were we assigned the least active or the swiftest abode, and does the god make everything move around us, or does he move us around?

Ptolemy's *Planetary Hypotheses*. Ptolemy's personal observations, as recorded in the *Syntaxis*, are spread over 17 years, from A.D. 124 to 141. The tables in the *Syntaxis* are scattered throughout. Later, they were collected and published separately by Ptolemy in his *Handy Tables*. Closely related to the *Handy Tables* was his popularizing work on the *Planetary Hypotheses*, in two Books. Only the first part of Book 1 is preserved in the Greek original. Nevertheless, the rest of the *Planetary Hypotheses* is not lost. For in certain instances, where a Greek original has disappeared, a translation into Arabic survives. This is what happened in the case of Ptolemy's *Planetary Hypotheses*. Although only the first part of Book 1 exists in Greek now, when the entire work was still available, it was translated into Arabic, and then from Arabic into Hebrew. When Ptolemy's collected works were printed in the early part of this century, by an oversight the *Planetary Hypotheses* lacked Book 1's second part. This missing piece was recently recovered from the Arabic and Hebrew versions, and translated into English.

Ptolemy's Dismemberment of Aristotle's Concatenated Spheres. In the recently recovered second part of Book 1 of his *Planetary Hypotheses*, Ptolemy says: "Each of the planets has one free motion, the other is determined of necessity."[12]

The necessary motion takes place from east to west, in unison with the universal daily rotation. For Ptolemy followed Aristotle's view that the daily rotation of the outermost sphere - the sphere of the stars - was made necessary by its ceaseless yearning for the Unmoved Mover beyond the stars.

On the other hand, Ptolemy discarded the rest of Aristotle's celestial machinery. For in Book 2, Chapter 6, of his *Planetary Hypotheses*, Ptolemy expressed disbelief that "there is anything in Nature that would be meaningless and useless."

> The same foolishness and absurdity are present also in spheres which counteract one another, to say nothing about the immense increase in the numbers [of the spheres]. For they take up a big space in the aether, and are not needed for the motions observed in the planets.[13]

Why do the Planets Move? Ptolemy got rid of Aristotle's interlocking spheres mechanically transmitting motion from planet to planet below the stars. Then why do the planets move, according to Ptolemy? One of their motions was necessary - the participation in the daily rotation. But their other motion was free. This was comparable to the free motion of birds, Ptolemy thought:

> Take the birds which we see as an example of the motion of the bodies observable in the heavens...When the birds move in one of their characteristic motions, the start of that motion is in a vital force inherent in them. This vital force gives rise to an impulse, which then spreads into the muscles. It then moves from the muscles, for example, into the feet or claws or wings, and there it stops.

Any contact between birds in flight would not promote motion, but would impede it, Ptolemy continues:

> There is no compelling reason to assume that the movements of all or most birds occur through their contact of one with another. On the contrary, it is necessary to postulate that they do not touch one another, unless we wanted them to interfere with one another.

Ptolemy then likened the flight of a flock of birds to the motion of the planets:

> In the same way, we may conceive the situation in the heavenly bodies. We may believe that every planet has its appropriate vital force, and moves itself. To the bodies linked with it by their nature, it transmits a motion. This motion begins in the closest body, and extends to the body attached thereto. Thus, the motion passes first to the epicycle, then to the eccentric, and then to the circle centered on the midpoint of the universe.

The vital force present in the planets and birds exists also in human beings. Just as the planet's vital force activates the epicycle, eccentric, and concentric, so the power of our brain drives our other bodily organs:

> The communicated movement differs, however, in the various places where it is received. In the same way, in ourselves the force of thought is not equal to the strength of the impulse. The power of this impulse is not equal to what acts on the muscles, nor is this equal to the force which activates the feet.[14]

The Arabic Translation of Ptolemy's *Planetary Hypotheses*. The Arabic translation of Ptolemy's *Planetary Hypotheses* was revised in the ninth century by an outstanding mathematician and astronomer. His version may be compared with the surviving Greek original in Book 1, part 1. In addition, while the whole Greek text was still available, a passage in Book 2 was quoted by a commentator on Aristotle's treatise *On the Heavens*:

> We must heed the best of the astronomers, Ptolemy, who says in Book 2 of his *Planetary Hypotheses*:
>
> It is therefore more reasonable that each of the planets is a source of motion . For this is its force and effect: a uniform and circular motion around its own center and in its own place. For it is proper for the planet itself to originate the motion which it imparts to the structures surrounding it.[15]

The commentator's quotation demonstrates the basic soundness of the revised Arabic translation. Here and there, however, its meaning is not entirely clear. Those difficulties would disappear if the missing parts of the Greek text of Ptolemy's *Planetary Hypotheses* were ever recovered.

The Arabic translation, as revised in the ninth century, referred to Ptolemy's masterpiece as the *Syntaxis*. The Hebrew version, five centuries later, shifted to the Arabo-Greek misnomer.[16]

Notes

5. *Babylonian Talmud*, Seder Mo'ed, Eruvin 56a, tr. I.W. Slotki (London, 1938), p. 394; Shabbath 156a, tr. H. Freedman, p. 799.

6. Plutarch, *The Face on the Moon,* ¶ *923.*

7. Dercyllides, quoted by Theon of Smyrna, *Mathematical Aids for Reading Plato*, Chapter 41 (ed. T.H.Martin, Paris, 1849, p. 328).

8. Joseph Needham, *Science and Civilisation in China,* III (Cambridge, England, 1959), 224.

9. *The Aryabhatiya of Aryabhata,* tr. Walter Eugene Clark (Chicago, 1930), p. 64.

10. In Chapter 13 of his *Panchasiddhantika*, Varaha-Mihira mentioned another argument: if the earth rotated in a day, flags and similar things would constantly stream toward the west, owing to the quickness of the rotation; tr. G. Thibaut (Benares, 1889; reprinted, Lahore, 1930), p. 82.

11. *Alberuni's India,* tr. Edward C. Sachau (London, 1910; reprinted, Lahore, 1962, under the authority of the government of Pakistan), I, 371-373.

12. Bernard R. Goldstein, "The Arabic Version of Ptolemy's *Planetary Hypotheses*," in *Transactions of the American Philosophical Society*, 1967, vol. 54, part 4, p. 6.

13. Ptolemy, *Opera astronomica minora,* ed. J.L.Heiberg (Leipzig, 1907), p. 118/1-2,20-25.

14. *Ibid.*, p. 119/21-23, 25-31; p. 120/1-20.

15. Simplicius (n. 3), p. 456/22-27; Ptolemy, *Opera astronomica minora*, p. 131/9-15.

16. Goldstein (n.12), p.7, n.2; p.8, n. 12.

CHAPTER 3

ECCENTRICS AND EPICYCLES, REAL OR UNREAL?

The Athenian Academy founded by Plato lasted nearly a thousand years. Toward the end of its life, its most important leader was the famous and influential Neoplatonist, Proclus (c.410-485). He was aware that Plato, like his pupil Aristotle, never mentioned eccentrics or epicyles, these having been first devised after Aristotle's death. Yet Proclus understood full well that the eccentric-epicyclic astronomy matched what was observed in the heavens much better than did Plato's dialogs and Aristotle's treatises. Hence Proclus faced a cruel dilemma. On the one hand, he could not dismiss the eccentric-epicyclic astronomy, whose technical value he recognized. On the other hand, he could not accept eccentrics and epicycles as real. For according to the Platonic philosophy, of which he was an outstanding exponent, all celestial motion must be centered on the earth, the immovable midpoint of the universe. But an eccentric, by definition, was not centered on the earth. So, too, an epicycle was centered on the circumference of a deferent, which might itself be an eccentric. The eccentrics and epicycles, though technically valuable, were philosophically unacceptable to Proclus. How could he escape from this dilemma?

Commentary on Plato's *Timaeus*. By the age of twenty-eight Proclus wrote, among other things, a commentary on Plato's dialog, the *Timaeus*. Of all his commentaries on Plato's dialogs, only five have survived, in whole or in part. Of these, his favorite was his commentary on the *Timaeus*. In it he concluded that the epicyclic-eccentric astronomy was ingenious but unassimilable. In themselves, what were the eccentrics and epicycles? This question was discussed in *Proclus' Commentary on Plato's Timaeus*:

> Hence it is now evident that, according to Plato, all the spheres are concentric and have the same center as the universe. On the other hand, the apparently irregular motions of the seven planets are caused by the way in which they change their movements in every possible manner, speeding up, slowing down, moving ahead, rising, and approaching the earth, without the artificial epicycles. For no such thing was mentioned by Plato, and Nature requires an intermediary

> everywhere. Now between all uniform and orderly things, on the one hand, and the nonuniform and disorderly, on the other hand, is that which is both irregular and orderly. Such is the nature of the planetary motion. Against its irregularity it balances a perpetually invariable pattern of speed and slowness, and movement toward or away from the same goal.
>
> But some people have used certain epicycles or eccentrics. They hypothesize uniform motions, in order to be able to compute the motions by combinations in which the epicycles and eccentrics move, with the planets on them. The idea is beautiful and suitable for logical minds. But it is alien to the nature of the universe, which only Plato apprehended.

Proclus scoffed at the epicycles and eccentrics again in a later passage of his *Commentary on Plato's Timaeus:*

> The spheres on which the planets move are now called by Plato "circles," but not "epicycles." For he mentioned these nowhere, just as he did not mention the eccentrics among the circles. For it would be ridiculous to make some little circles move on every sphere in the direction opposite to its, or make them parts of the sphere, or of another substance, or eccentric spheres embracing the center without moving around it. For this undermines the common axiom of the physicists that every uniform motion is either around the center of the universe or away from it or toward it....Plato never moves the planets in any different way, nor does he need such devices, unworthy of divine beings.[17]

Proclus' Commentary on Plato's *Republic*. Proclus discussed the eccentrics and epicycles, as well as the counteracting spheres, in his *Commentary on Plato's Republic.* There he noted with approval that Plato

> has no room for the counteracting spheres or the eccentrics or the epicycles or, in general, for these mechanized hypotheses. Rightly so. For those who employ the hypotheses of the eccentrics make these spheres move around a center other than the center of the fixed stars [which is the center of the universe]....For if moving around the center of the universe is best, how is it that the heavenly spheres use other centers? But if it is not best, how is it that around the center of the universe there rotates the sphere of the stars, which controls the inner spheres and makes them turn?

After disposing of the eccentrics Proclus tackled the epicycles:

> Moreover, the hypothesis employing the epicycles is utterly ridiculous....The epicycles introduce the concept of mechanics to a great extent. For suppose that we conceived some small circles, with the planets on them. To make circles, instead of

> bodies, revolve in the heavens is plain foolishness.

As an alternative to mounting the planets on two-dimensional circles, Proclus considered the possibility of placing the planets on three-dimensional spherical bodies:

> Or suppose some spherical bodies are attached to the [deferent] spheres and moved by them, with the planets on the spherical bodies. This is even more impossible and more fictitious. It does not differ at all from having the planets transported on some vehicles, as though they could not move themselves, and had to be carried by others. It is laughable for those [who] want to keep the motion uniform to do so by doubling the things in motion.

"To Save the Phenomena." Proclus then proceeded to distinguish between two approaches to the problem of planetary motion. Those who came from the side of philosophy believed that the heavenly motions were thoroughly rational. But those who were primarily mathematicians wanted to "save the phenomena." This aim had become a dominant theme in Greek astronomy. A normal man under ordinary circumstances walks forward, not backward, to reach his goal. Then why does a planet, which was believed to be divine and loftier than man, sometimes move backward? The mathematicians were convinced that a planet always moved forward. To us on earth, it merely seems to be moving backward. Hence the Greek mathematicians devised models of planetary motion, in which the planet always moved forward, but sometimes seemed to be moving backward. Such models were designed to "save the phenomena." As Proclus put it,

> Those who approach [this problem] from the mathematical side want to adopt such hypotheses as eccentrics and epicycles "in order to save the phenomena." But according to those who look to the Muse of philosophy, it is necessary to be on guard that nothing is done irrationally, especially [in heaven,] where nothing is purposeless or accidental, but everything is in accordance with reason.

"To Explain the Most with the Least." Proclus' next objection to the eccentric-epicyclic astronomy recalled the familiar exhortation to "explain the most with the least":

> The Pythagoreans had a maxim...that the restoration of the observed nonuniformity of the heavenly bodies to uniformity and orderliness requires the fewest and simplest hypotheses. But those who use the counteracting spheres are far from doing so. They take numerous hypotheses for the phenomena, manufacture countless spheres, and fabricate a manifold cosmos in order to devise a simple arrangement for a single planet.

Last of all, Proclus objected to the incompleteness of the eccentric-epicyclic astronomy:

> Moreover, those who came afterwards refuted these hypotheses as being incapable of saving all the phenomena, and not explaining adequately the phenomena they do save.[18]

Proclus' *Hypotyposis*. When the Christians became dominant in Athens, they made life difficult for those who chose to cling to the ancient Greek gods. Proclus had to flee, and escaped to the interior of Asia Minor (now Turkey). There, as a childless bachelor, he lived with a friend who was eager to understand what the famous Greek astronomers had contrived. Proclus promised to satisfy his friend's desire when he had the opportunity. The following year, after he returned to Athens, he wrote an *Outline (Hypotyposis) of the Astronomical Hypotheses.*

Proclus himself was not a creative or innovative astronomer. But he was a first-rate teacher of astronomy at the introductory level. In his *Hypotyposis* he summarized the eccentric-epicyclic system compactly and competently. In particular, he had a thorough knowledge of the *Syntaxis.* He referred to its author, Ptolemy, as "wonderful," and to his treatment of astronomical topics as "very clear" and "outstanding" (Ch. V,1, beginning; Ch. V,5, beginning; Ch. VI, end).

The Eccentrics and Epicycles in Proclus' *Hypotyposis*. At the very outset Proclus expressed his confidence that his *Hypotyposis* would make his friend too realize that it was a refutation of the professional astronomers' hypotheses. Likewise, in the seventh and last chapter of the *Hypotyposis,* after emphasizing ten main features of the astronomical hypotheses, Proclus asked:

> What shall we say concerning the eccentrics and epicycles, about which the professional astronomers keep talking? Are these merely imagined, or do they also have physical reality in their spheres, in which they are fastened? For if they are merely imagined, the astronomers unwittingly shifted from physical bodies to mathematical concepts, and derived the causes of natural motions from entities that do not exist in nature.

Proclus' Objections to the Astronomers' Mathematical Constructions. Forward motion at a constant speed - a theoretical requirement, according to Proclus - was not exhibited by the interconnected eccentrics, epicycles, and deferents:

> I shall also add that the astronomers would be illogical in describing the motions. For, the motions are in accordance with our concepts. Hence, the heavenly bodies imagined on

> the eccentrics and epicycles do not really move nonuniformly. And if the eccentrics and epicycles also exist physically, their connection with the [deferent] spheres in which these circles exist is obliterated by the astronomers. They make these circles move independently, and the [deferent] spheres independently. The circles do not even move like one another, but in opposite directions.

Proclus objected vigorously to the astronomers' mathematical constructions for the planets' motions in latitude:

> Moreover, the distances of these circles from one another are confused, since sometimes they come together and lie in one plane, while at other times they diverge and intersect one another. Hence there will be all sorts of divisions and coalescences and separations of the heavenly bodies.

The astronomical hypotheses were unsatisfactory because they were arbitrary:

> Furthermore, the treatment of these mechanized hypotheses also seems haphazard. For in each hypothesis why is the eccentric as it is, either stationary or movable? Why is the epicycle as it is, with the planet being moved either eastward or westward? And what are the reasons for those planes and divergences? I mean the real reasons, those which definitely put an end to all travail of the mind when it understands them. No answer of any sort is forthcoming.

Geometers started with hypotheses, and drew conclusions from them. But the astronomers reverse this process:

> Instead, actually proceeding backward, the astronomers do not draw systematic conclusions from the hypotheses, as the other disciplines do. On the contrary, from the conclusions they try to form the hypotheses, from which they should have inferred these conclusions. And in the process they do not even appear to have provided what was possible.

Proclus' Compromise. As we have just seen, Proclus objected strenuously to the astronomers' mathematical devices. These were, however, more suitable in his judgment than all other competing devices for the purpose of computing the motions of the planets:

> Yet this much should be known, that these are simpler than all the [other] hypotheses, and more suitable for divine bodies. These hypotheses have been devised to find out how the planets move, since they really move as they seem to move. The purpose is to bring within reach the measure of their motions.[19]

On this conciliatory note, Proclus ended his *Hypotyposis*: to compute the heavenly motions, the eccentrics and epicycles were serviceable, as mathematical concepts, but not as physical bodies.

The Spanish Muslims: Ibn Tufayl and Al-Bitruji. In the seventh century the Muslim movement burst out of Arabia and conquered most of the eastern provinces of the Later Roman or Byzantine empire. Islam then swept westward from Egypt and overran the southern shore of the Mediterranean Sea. Early in the eighth century the Muslims crossed over from North Africa into Spain, which they controlled for centuries. The Spanish Muslims revered Aristotle as the greatest of the philosophers. Finding the eccentric-epicyclic astronomy in conflict with Aristotle's astronomical teachings, they searched for an alternative to the Ptolemaic system.

Abu Bakr Ibn Tufayl (c.1110-1185), who was known in Latin as Abubacer, was said by a pupil

> to have been inspired with an astronomy, whose principles differed from the two principles adopted by Ptolemy. Neither the eccentric nor the epicycle was accepted by Tufayl. Thus he saves all the motions, so that nothing impossible is involved in his astronomy, about which he promised to write a book.[20]

If Ibn Tufayl ever wrote that book, it has not come down to us. But his pupil Al-Bitruji, known in Latin as Alpetragius, did write a book about astronomy. This was translated into Latin by Michael Scot in 1217. Al-Bitruji explained that in Ptolemy

> the planet moves on an epicycle, whose center revolves on an eccentric, whose center is outside the center of the universe.

Ptolemy's eccentric deferents required

> a vacuum in which those eccentrics move....Then one region is left empty, and another region is filled up. All this is disgraceful, far from the truth, and different from the reality of the heavens.[21]

Al-Bitruji was persuaded to

> reject those eccentric spheres and epicycles, of which there is a large number among those of the older practitioners of this science who postulated them.[22]

In opposition to those astronomers, remote and recent, who invoked epicycles and eccentrics, Al-Bitruji adhered firmly to the principle that "the center of all [the motions] is the center of the universe."[23]

Ibn Rushd. Ibn Tufayl was a physician by profession, and served the sultan of Morocco and Andalusia in that capacity. Another physician introduced to the sultan by Ibn Tufayl was Ibn Rushd (1126-1198). He was convinced that Aristotle

> comprehended the whole truth, and by the whole of the truth I mean that quantity which human nature - insofar as it is human - is capable of grasping.[24]

A famous characterization valued Aristotle as "the master of those who know." But he certainly was not the master of those who communicate. Hence, a need was felt to expound his views. Such an exposition took the form of a commentary. Of all the Islamic commentators on Aristotle, the most celebrated was Ibn Rushd. His name was latinized as Averroes when his writings were translated into that language. Just as Aristotle came to be known as "the Philosopher" without further ado, so Ibn Rushd was often designated "the Commentator."

Ibn Rushd's Commentary on Aristotle's *Metaphysics*. One of Aristotle's most influential works was his *Metaphysics,* which was divided into fourteen Books. These were often identified by letters of the Greek alphabet. To an ancient Greek, a letter of his alphabet could stand for a number as well as for a sound. The Greeks lacked a numerical notation apart from their alphabet. In Book Lambda, or Book 12, of his *Metaphysics* Aristotle dealt briefly with astronomy and its history in Chapter 8. There he gave the number of the celestial spheres in his universe as 55. They all shared the same center, which was the center of the earth. This neat and orderly arrangement of 55 concentric spheres jibed with the basic structure of Aristotle's philosophy. In his cosmos there were neither eccentrics nor epicycles, devices introduced into astronomy long after Aristotle's death.

Some 1500 years later, however, the situation was entirely different. When Ibn Rushd in his old age wrote his *Commentary on Aristotle's Metaphysics,* the eccentrics and epicycles had long been part of the mathematical astronomical establishment. Since eccentrics and epicycles were incompatible with the Philosopher's system, Ibn Rushd repeatedly made them the target of fierce onslaughts.

Ibn Rushd's Metaphysics Commentary, 12, 2, 4, 45. Before the invention of printing with movable type in the fifteenth century, it was often hard to locate a desired passage in a lengthy manuscript. To make it easier for the reader to find his way in Ibn Rushd's voluminous *Commentary,* the Latin translator divided each of its Books into sections and subsections. The longest such section was called a summa. Each summa was in turn split up into chapters. Each chapter was further subdivided into numbered comments. In Book 12, summa 2, chapter 4, comment 45, Ibn Rushd blasted the eccentrics and epicycles with exceptional vigor:

> Talking about an eccentric or an epicycle is unnatural. It is altogether impossible for an epicycle to exist. For, a body moving in a circle travels only around the center of the universe, since the moving body determines the center. If there were a circular motion not related to this center, outside this center there would be another center. In addition to this earth, therefore, there is another earth, and that is impossible....

> Similar reasoning perhaps applies to the eccentric assumed by Ptolemy. For if there were several centers, there would be several heavy bodies outside the place of the earth. In that case the center would not be unique, it would possess latitude, and be divided. All these consequences are impossible. Moreover, if there were eccentrics, among the heavenly bodies there would have to be found unnecessary bodies serving no purpose except to fill up empty space, as in the bodies of animals. On the other hand, nothing of what we see in the motions of the stars compels us to admit the existence of an epicycle or eccentric....
>
> Since the epicycle and eccentric are impossible, we must therefore think our way back again to the true astronomy, which is based on natural foundations....When I was young, I hoped to complete this research. But now that I am old, I have abandoned this hope. Nevertheless, this discussion may perhaps induce someone else to carry out this investigation. For, the astronomy of the present time is nonexistent as far as reality is concerned. But it is in agreement with the computations, not with reality.[25]

The Search for an Alternative to Eccentrics and Epicycles. Earlier in his career, in his commentary on Aristotle's *Heavens*, with reference to lunar eclipses, Ibn Rushd expressed the hope that

> It may be possible to find an astronomy in agreement with the observations of the moon, without an eccentric sphere (Book 2, summa 2, question 4, comment 32).[26]

In Book 2, summa 2, question 5, comment 35, Ibn Rushd extended the attack to epicycles:

> In mathematics there is no apparent reason to believe in epicycles or eccentrics. For, those spheres assumed by the astronomers are antecedent constructs, from which the phenomena perceived by the senses follow. In this context there is no proof that those antecedent constructs necessarily follow from those phenomena.[27]

The thrust of Ibn Rushd's reasoning here is perfectly plain. Grant that what we observe in the heavens is deduced from an epicyclic-eccentric system as an antecedent construct. That deduction does not in the least prove that the epicyclic-eccentric system is true. Such a deductive proof would be valid only if the observed phenomena could be deduced from no other antecedent constructs than an epicyclic-eccentric system. If the observations can be deduced equally well from any alternatives to the epicyclic-eccentric system, then the relative merits of the rival antecedent systems must be weighed in the balance. To ignore the possible competitors of the eccentrics and epicycles would be wrong. For, such an oversight would leave the door open to the claim that the eccentrics and epicycles must exist because they offer the only way to explain the observed phenomena. Opening this door would admit the fallacy denounced by logicians as "affirming the consequent." Eager to block the acceptance of this common fallacy, Ibn Rushd recognized that he had to look for a sound alternative to the eccentrics and epicycles.

Why Ibn Rushd went Astray. This alternative he hoped to find in Aristotle. But here he was hampered by his ignorance of Greek. Ibn Rushd did not base his Aristotle commentaries on the original Greek text. Because the Commentator did not know the language of Aristotle, he had to rely on translations of the Philosopher into Arabic. Those translators did not share our ideal of what a translation should be: reproduce a text as faithfully as possible in another language. On the contrary, they felt free to insert interpolations without warning readers that they were doing so. Hence Ibn Rushd failed to realize that what he had before his eyes was something other than the genuine text of Aristotle. Unaware that he was working with an adulterated text, and being unfamiliar with Greek, Ibn Rushd was in no position to separate Aristotle's own thinking from the Arabic interpolations. That is why Ibn Rushd mistakenly attributed to Aristotle some non-Aristotelian ideas.

These non-Aristotelian ideas, misattributed to Aristotle by Ibn Rushd, included a sort of spiral motion for the planets. This supposed planetary spiral motion goes back to a late Greek commentary on Ptolemy. Finding it in an Arabic "translation" of Aristotle, Ibn Rushd erroneously imagined that it was known to Aristotle, nearly 500 years before Ptolemy. Hence Ibn Rushd anachronistically charged that

> Ptolemy did not understand why his predecessors had to accept the spiral motion. The reason is that the epicycle and eccentric are impossible.[28]

Actually Ptolemy's predecessors, including Aristotle, did not accept the spiral motion. They did not accept it because they never heard of it. The spiral motion was not the reason why Aristotle kept quiet about the epicycle and eccentric. Aristotle said nothing about epicycles and eccentrics because he never heard of them. They were rejected as impossible, not by Aristotle himself, but by that staunch Aristotelian, Ibn Rushd. In his *Commentary on the Heavens* (Book 2, question 5, comment 62) Ibn Rushd expressed his conviction that

> the motions needed for these phenomena have not yet been demonstrated in this science. For, the motions postulated by

> Ptolemy are based on two foundations which are inappropriate in natural science, namely, the eccentrics and epicycles, both of which are false.[29]

Maimonides' Acceptance of Eccentrics. To get rid of the Ptolemaic eccentric and epicycle was one aim of the Aristotelian movement that arose in Spain during the twelfth century. Its loudest voice belonged to Ibn Rushd. Other Muslim intellectuals joined the chorus led by the Commentator. Their music appealed to a Jew who became his people's foremost philosopher in the Middle Ages. Maimonides (1135/1138-1204) grew up in Muslim Spain and wrote mainly in Arabic. This was the original language of his most famous work, the *Guide of the Perplexed,* the outstanding achievement of Jewish medieval philosophy. Completed shortly before his death, the *Guide* was promptly translated into Hebrew by two of Maimonides' co-religionists. In a letter to one of these translators, Maimonides expressed his admiration for Aristotle and Ibn Rushd:

> The works of Aristotle are the roots and foundations of all works on the sciences. But they cannot be understood except with the help of commentaries...[including] those by Ibn Rushd.[30]

Yet Ibn Rushd's commentaries on Aristotle were not yet available to Maimonides when he was writing the *Guide.* By contrast to Ibn Rushd, in the *Guide* Maimonides did not exclude the eccentrics. Thus Maimonides discussed (I, 72) "the sphere of the outermost heaven with everything that is within it":

> The sphere in question as a whole is composed of the heavens, the four elements, and what is compounded of the latter. In that sphere there is absolutely no vacuum; it is solid and filled up. Its center is the sphere of the earth.

Between the earth and the outermost heaven, Maimonides located

> many spheres, one contained within the other, with no hollows between them and no vacuum in any way whatever. For they are perfectly spherical and cling to each other, all of them moving in a circular uniform motion, in no part of which is there acceleration or deceleration.

Whereas in the cosmos of Aristotle and Ibn Rushd all the heavenly spheres had the same center, Maimonides' spheres

> have different centers. The center of some of them is identical with the center of the world, while the center of others is eccentric to the center of the world.[31]

Thus, for Maimonides the eccentrics exist. But he is not equally certain about the epicycles: "it is also a matter of speculation whether there are epicycles."[32]

Maimonides' Rejection of Eccentrics and Epicycles. Later on in the *Guide*, however, Maimonides' uncertainty about epicycles came to an abrupt end. In order to explain what is observed in the heavens, he asserts (II, 24):

> everything depends on two principles; either that of the epicycles or that of the eccentric spheres or on both of them....Both these principles are entirely outside the bounds of reasoning and opposed to all that has been made clear in natural science.[33]

In this connection Maimonides repeats the statement that "the existence of epicycles is impossible." He quotes this rejection of the epicycles from a "discourse on astronomy" by Ibn Bajja, the founder of the Aristotelian movement in Muslim Spain, who had died at just about the time when Maimonides was born. Maimonides describes his "discourse on astronomy" as extant. Unfortunately, it has since disappeared. Nevertheless, we know a little about this discourse by Ibn Bajja (or Avempace, as his name was latinized). Both of Ibn Bajja's reasons for declaring the epicycle impossible were summarized by Maimonides. First,

> The revolution of the epicycles is not around the center of the world. Now it is a fundamental principle of this world that there are three motions: a motion from the midmost point of the world, a motion toward that point, and a motion around that point. But if an epicycle existed, its motion would be neither from that point nor toward it nor around it.[34]

Besides ruling out the epicycle as excluded from Aristotle's three kinds of motion, Ibn Bajja condemned it for conflicting with an axiom of Aristotle's physics. By its very definition, the epicycle revolved around a center that moved along the circumference of its deferent. However,

> it is one of the preliminary assumptions of Aristotle in natural science that there must necessarily be some immobile thing around which circular motion takes place. Hence it is necessary that the earth should be immobile. Now if epicycles exist, theirs would be a circular motion that would not revolve around an immobile thing.

For these two reasons Ibn Bajja denounced the epicycles as impossible. Maimonides heard, however, that Ibn Bajja "had invented an astronomical system in which no epicycles figured, but only eccentric circles." To this system Maimonides objected that

> eccentricity also necessitates going outside the limits posed by the principles established by Aristotle, those principles to which nothing can be added....In the case of eccentricity, we likewise find that the circular motion of the spheres does not take place around the midmost point of the world, but around

> an imaginary point that is other than the center of the world. Accordingly, that motion is likewise not a motion taking place around an immobile thing.[35]

In Aristotle's cosmos there was nothing immobile outside the earth. Everything outside it participated in the universe's constant motion. By definition, an eccentric's center would be outside the earth's center. Then the eccentric's center would be moving, not stationary. Hence, the eccentric would revolve around a moving point. But, according to Aristotle, the center of a circular motion must be at rest. Therefore, in the thinking of Maimonides, Ibn Bajja's eccentrics were just as impossible as the epicycles.

Maimonides Changes his Mind. Yet, earlier in the *Guide*, as we saw above, Maimonides accepted eccentrics without any qualms. Moreover, at that time (I, 72) he regarded it as a "matter of speculation whether there are epicycles." But after reading Ibn Bajja's discourse on astronomy, Maimonides changed his mind: in the *Guide*, II, 24, he rejected both eccentrics and epicyles.

Maimonides was extremely sensitive to "the contradictory statements to be found in any book." In the Introduction to the *Guide* he analyzed seven reasons for such contradictions. The second reason was that

> the author of a particular book has adopted a certain opinion that he later rejects; both his original and later statements are retained in the book.

Maimonides was acutely aware of this defect in the books he read. But was he equally aware of this defect in his own *Guide*? In I, 72, he accepted eccentrics and was doubtful about epicycles. Later on, in II, 24, he rejected both eccentrics and epicycles. Yet he did not then go back to I, 72, to make it consistent with II, 24.

Aristotle's Physics Incompatible with Ptolemy's Astronomy. This inconsistency, however, is as nothing compared to the basic incompatibility between Aristotelian physics and Ptolemaic astronomy:

> If what Aristotle has stated with regard to natural science is true, there are no epicycles or eccentric circles, and everything revolves round the center of the earth. But in that case how can the various motions of the stars come about? Is it in any way possible that motion should be on the one hand circular, uniform, and perfect, and that on the other hand the things that are observable should be observed in consequence of it, unless this be accounted for by making use of one of the two principles [epicycles and eccentrics] or of both of them?

Eclipses of the moon were predicted accurately on the basis of an eccentric-epicyclic lunar theory. Furthermore, Maimonides asks,

> how can one conceive the retrogradation of a star...without

> assuming the existence of an epicycle? On the other hand, how can one imagine...a motion around a center that is not immobile? This is the true perplexity.

How was this perplexity resolved by the author of the *Guide of the Perplexed?*

> All this does not affect the astronomer. For his purpose is not to tell us in which way the spheres truly are, but to posit an astronomical system in which it would be possible for the motions to be circular and uniform and to correspond to what is apprehended through sight, regardless of whether or not things are thus in fact.[36]

What is apprehended through sight includes the retrogradation of a planet. This is seen moving eastward in its orbit most of the time. But for a short while it pursues the opposite, or retrograde, direction. This apparent retrogradation of a planet was handily explained by the combination of an epicycle with a deferent. For while the planet traverses the arc of the epicycle that lies within the deferent, to an observer located at the center of the deferent the planet seems to be going backward. Meanwhile the center of the epicycle moves steadily forward along the circumference of the deferent. As a result, the planet moves around a center that is not immobile, but is itself in constant motion. The deferent-epicycle model accounted for the planetary retrogradations, but conflicted with the Aristotelian dogma that the center of a circular motion must itself have no motion.

Terrestrial Physics Versus Celestial Astronomy. A group of thinkers opposed by Maimonides maintained that

> Everything that may be imagined is an admissible notion for the intellect. For instance, it is admissible from the point of view of intellect that it should come about that the sphere of the earth should turn into a heaven endowed with circular motion.[37]

To Maimonides, the idea of the earth as a revolving heavenly body seemed reprehensible. Little did he imagine that some three hundred years later the motion of the earth would be demonstrated incontrovertibly. He himself, however, was perfectly satisfied with Aristotle's earthly physics:

> Everything that Aristotle has said about all that exists from beneath the sphere of the moon to the center of the earth is indubitably correct.

But Aristotle's astronomy was another story:

> Everything that Aristotle expounds with regard to the sphere of the moon and that which is above it is, except for certain things, something analogous to guessing and conjecturing.[38]

Since the peerless Aristotle produced an astronomy that was merely conjectural, what could be expected of that science?

> Regarding all that is in the heavens, man grasps nothing but a small measure of what is mathematical.

Man is able to know

> what is beneath the heavens, for that is his world and his dwelling-place, in which he has been placed and of which he himself is a part.

On the other hand,

> it is impossible for us to accede to the points starting from which conclusions may be drawn about the heavens; for the latter are too far away from us and too high in place and in rank...Let us then stop at a point that is within our capacity, and let us give over the things that cannot be grasped by reasoning.[39]

For Maimonides, then, the conflict between terrestrial physics and celestial astronomy ended in the elevation of physics and downgrading of astronomy: physics remained a pathway to the truth, whereas astronomy lay beyond human reach.

What science has achieved since Maimonides' time proves how wrong he was. In his day it may have been true that "regarding all that is in the heavens, man grasps nothing but a small measure of what is mathematical." That small measure has been greatly enlarged by the computer. To Maimonides the heavens seemed "too far away from us and too high" to be understood. To Neil Alden Armstrong, who on 20 July 1969 became the first man to step down to the surface of the moon, that heavenly body was neither too far away nor too high.

Theoretical Astronomy Versus Mathematical Astronomy. Before the age of the computer and the astronaut, however, Maimonides' judgment regarding mankind's incapacity to grasp astronomy was widely accepted. After his *Guide* was translated into Latin, its impact on Christendom was swift and pervasive. The outlook of many people had long been dominated by the preaching that humanity was helpless and unable to cope. To the depressing list of what was unattainable, astronomy was now added.

Ibn Rushd's point of view was quite different. He did not seek to dissuade people from studying astronomy by proclaiming that it rose far beyond their reach. He was convinced that the astronomy of his time was wrong in its basic concepts. Yet its computations could be useful. The Latin translation of his *Commentary on Aristotle's Metaphysics* carried his message far and wide, as we saw above at n. 25:

> The astronomy of the present time is non-existent as far as

> reality is concerned. But it is in agreement with the computations, not with reality.

Ibn Rushd's verdict split astronomy into two parts. Theoretical astronomy (of the Ptolemaic variety) was out of touch with reality. Mathematical or computational astronomy was dependable, even though its foundations were false.

Thomas Aquinas. With appropriate modifications, this was the position taken by the most highly esteemed Christian theologian of the Middle Ages, Thomas Aquinas (c. 1225-1274). His family's name denoted association with Aquino, a locality in central Italy. In his most important work, the *Summary of Theology (Summa theologiae),* Aquinas stated:

> an astronomical argument about eccentric and epicyclic motions is put forward on the ground that by this hypothesis one can show how celestial movements appear as they do to observation. Such an argument is not fully conclusive, since an explanation might be possible even on another hypothesis.[40]

In his *Commentary on Aristotle's Heavens* (Book 2, Chapter 12, Lectio 17) Aquinas also treated eccentrics and epicycles as acceptable in the short run, but not in the final analysis:

> The assumptions adopted by them [the astronomers] are not necessarily true. For even though the phenomena would be explained by the adoption of such assumptions, nevertheless it must not be said that these assumptions are true. For, the heavenly phenomena may perhaps be explained in some other way not yet understood by mankind.[41]

Unlike the Commentator and other Aristotelians, Aquinas did not deny the physical existence of epicycles and eccentrics. He admitted their usefulness in astronomy. But, regarding them as only provisionally valid, he refused to put the stamp of "ultimate truth" on them. Thus, he followed the Commentator in avoiding the fallacy of "affirming the consequent."

Astronomy, like other natural sciences, does not deal with ultimate truths. These it leaves to theologians who are convinced that such final verities are accessible to themselves. For science, on the other hand, the goal is rather ever closer approximation to the truth. The history of science is full of examples of discarded ideas. These may have seemed suitable for a time, only to be replaced by more acceptable refinements. Just as the concentrics were ousted from mathematical astronomy by the more serviceable eccentrics and epicycles, these in turn were later dropped in favor of the elliptical planetary orbit.

Notes

17. Proclus, *Commentary on Plato's Timaeus*, 272A-B, 284C-D, ed. Diehl (Leipzig, 1903-1906), III, 96/13-32, 146/14-22, 146/30-147/2.

18. Proclus, *Commentary on Plato's Republic*, ed. Wilhelm Kroll (Leipzig, 1899-1901), II, 227/26-30, 228/28 - 229/2, 229/8-9, 10-21, 229/26 - 230/1, 230/2-9, 11-13.

19. *Procli Diadochi Hypotyposis astronomicarum positionum*, ed. Manitius (Leipzig, 1909), p. 236/15 - 238/27; reprinted, Stuttgart: Teubner, 1974.

20. *Al-Bitruji, De motibus celorum*, ed. Francis J. Carmody (Berkeley/Los Angeles, 1952), p. 78, III, 1; p. 76, III, 11.

21. *Ibid.*, p. 78, III, 25.

22. *Ibid.*, p. 90, VII, 17.

23. *Ibid.*, p. 91, III, 24.

24. *Dictionary of Scientific Biography* (New York, 1970-1980), XII,6.

25. *Aristotelis opera cum Averrois commentariis* (Venice, 1562-1574; reprint, Frankfurt am Main, 1962), VIII, 329vG-M.

26. *Ibid.*, V. 116rA.

27. *Ibid.*, V, 118vK.

28. *Ibid.*,VIII, 329vK; Ibn Rushd's commentary on Aristotle's *Metaphysics*, Book 12, summa 2, chapter 4, comment 45.

29. *Ibid.*, V, 140vL.

30. Moses Maimonides, *The Guide of the Perplexed*, tr. Shlomo Pines (Chicago, 1963), p. lix.

31. *Ibid.*, p. 184.

32. *Ibid.*, p. 185.

Notes

33. *Ibid.*, p. 322.

34. *Ibid.*, p. 323.

35. *Ibid.*

36. *Ibid.*, pp. 325-326.

37. *Ibid.*, p. 206.

38. *Ibid.*, pp. 319-320.

39. *Ibid.*, pp. 326-327.

40. Thomas Aquinas, *Summa theologiae*, Blackfriars ed., VI (London/New York, 1965), 106.

41. Thomas Aquinas, *Opera omnia* (Rome, 1886), III, 186-187.

Notes -- Chapter 4

42. *In Metaphysicen Aristotelis Quaestiones...Ioannis Buridani* (Paris, 1518; reprint, 1964), sig. AA1r,v.

43. *Ibid.*

44. Hubert Pruckner, *Studien zu den astrologischen Schriften des Heinrich von Langenstein* (Leipzig, 1933), *Tractatus contra astrologos coniunctionistas de eventibus futurorum*, I,2, p. 140/11-15.

45. *Ibid.*, I,8, p. 151/16-17.

See also p. 64.

CHAPTER 4
COPERNICUS AT THE UNIVERSITY OF CRACOW

While the battle over the existence or non-existence of eccentrics and epicycles was still raging, that hot question was handled by a highly popular lecturer at the University of Paris, Jean Buridan (c. 1295 - c. 1358). His *Questions on Aristotle's Metaphysics* was printed in Paris in 1518, and reprinted in Frankfurt am Main in 1964. In discussing Book 12 of Aristotle's *Metaphysics,* Buridan began Question 10 by asking "whether epicycles should be admitted among the heavenly bodies." In like manner he started Question 11 with the query "whether eccentric orbs should be admitted in heaven."[42] With regard to both epicycles and eccentrics, he commenced with the negative answer: they are equally inadmissible. His first argument for the negative side was "the authority of the Commentator [Ibn Rushd], who explicitly tries to disprove" epicycles and eccentrics. One of Ibn Rushd's arguments was summarized by Buridan as follows:

> Every heavenly motion must be circular, and therefore must be around something. The motion of an epicycle, however, would not be around anything. For, the epicycle would have no center other than an indivisible point around which to move. Such an indivisible point is not a natural center. In fact, it is nothing. Hence that motion is not around anything.

For Buridan as a loyal Aristotelian,

> The natural center in the universe is the earth itself. Hence an indivisible center is assumed only in the imagination.

Why were epicycles and eccentrics adopted "by Ptolemy and all modern astronomers"? According to Buridan, without these devices,

> there could be no explanation of what we see in the planets, especially their coming closer to the earth and receding farther from it. Yet these variations are obvious and important parts of our experience....Up to the present no way has been found, nor does it seem that it could be found, to explain...planetary phenomena except by assuming eccentrics or epicycles.

To this strategy of the astonomers, Ibn Rushd reacted as follows, according to Buridan's summary:

> This way of assuming or imagining epicycles and eccentrics is quite valid for computing and knowing the positions of the planets and their configurations with respect to one another and to us. Nothing more than this is sought by the

> astronomers. Accordingly, they are permitted to use such figments of the imagination even though these do not really exist.[43]

Henry of Hesse. In the generation after Buridan at the University of Paris, Henry of Hesse (also called "of Langenstein"; 1325-1397) was disturbed by the astrological predictions evoked by the conjunction of Mars and Saturn in March 1373. In his *Treatise against the Astrologers [Who Predict] Events in the Future on the Basis of Conjunctions* Henry of Hesse declared:

> For the purpose of saving the nonuniform phenomena in the motions of the heavenly bodies, the speculative astronomers have imagined eccentrics, epicycles...concerning which it is uncertain whether this is so, many astrologers and philosophers regarding it as impossible.[44]

Among those philosophers was Ibn Rushd, whose condemnation of epicycles was repeated by Henry of Hesse:

> Epicycles are nothing in reality, there is only doubt about them.[45]

Cracovian Uncertainty about Eccentrics and Epicycles. Ibn Rushd's attitude toward eccentrics and epicycles spread far and wide beyond the University of Paris. In particular, the University of Cracow, in what was then the capital of the kingdom of Poland, felt the influence of Ibn Rushd's follower, Buridan. Manuscript copies of many of Buridan's writings were acquired by the Jagellonian Library, named after the ruling Polish dynasty, which founded the university and its library. This also possessed a considerable number of different commentaries on Aristotle's *Metaphysics.* These commentators reported Ibn Rushd's and Buridan's opposition to eccentrics and epicycles, which were nevertheless used by the astronomers.

Copernicus in Cracow. This unsettled situation still prevailed when a freshman named Nicholas Copernicus (1473-1543) entered the University of Cracow in the winter semester of 1491. The subjects offered during his years there are known from university records that are still preserved. The names of the professors who taught those courses are also known. But the students' records are missing. Hence, we do not know who instructed Copernicus at Cracow, nor what grades he received. We do not even know how many years he remained in Cracow. One thing is certain: he did not stay long enough to earn a bachelor's degree, which normally required four years.

Copernicus the Canon. Copernicus' failure to acquire the baccalaureate degree does not show him to have been stupid or lazy. For he did not need that degree in the career he had in mind. His father, Nicholas Copernicus Sr., had died when Nicholas Jr. was only ten years old. The boy was lucky enough, however, to have an uncle who acted like a father toward him. This surrogate father, the brother of Nicholas Copernicus' mother, was a prosperous clergyman, who in 1489 became the bishop of Varmia. This diocese had a cathedral chapter of sixteen canons, each of whom received a substantial income for life. If one of these sixteen canons died, and Copernicus obtained the vacant canonry, his material wants would be satisfied for the rest of his life. But a canon of Varmia was expected to study at some recognized institution of higher learning for three years, and then return with an advanced degree. Copernicus' maternal uncle had studied canon law at the University of Bologna in Italy. The nephew planned to follow in his uncle's footsteps. No bachelor's degree was required for admission to the law curriculum at the University of Bologna. Hence Copernicus, like many other students at the University of Cracow, did not remain there long enough to acquire the baccalaureate degree.

Copernicus the Astronomer. Although we do not know exactly what courses Copernicus took at the University of Cracow, or with whom he studied, his own interest in astronomy coincided with the university's emphasis on the mathematical disciplines. The reputation enjoyed by Cracow in this field attracted many foreign students, including those whose mother tongue, like Copernicus', was German. The language of instruction in Cracow, as in all the other universities of Europe at that time, was Latin. We have so little information about Copernicus' early life that we are not sure where he learned Latin before entering the University of Cracow at the age of eighteen. It was there that he mastered the fundamentals of mathematics and astronomy, as he himself acknowledged. His recognition of his intellectual debt to the University of Cracow was publicized in a book printed in Cracow not long before his death. *(See Reading No. 7).*

Did Copernicus Develop his own Astronomy in Cracow? Copernicus' acknowledgment of his indebtedness to the University of Cracow was made public by a young author whom he had never met face to face. This youngster must have obtained the acknowledgment from one of the Cracow astronomers with whom Copernicus carried on a professional correspondence. For after leaving the University of Cracow, Copernicus had no further occasion to revisit that city. His letter to that Cracow astronomer has not been preserved. In it he evidently quoted the judgment that "to identify those from whom we have benefited is an act of courtesy and thoroughly honest modesty." Copernicus took this quotation from the ancient Roman writer, Pliny, the author of the famous *Natural History.* Book II of Pliny's *Natural History* deals with astronomy. Some of Pliny's expressions in Book II appealed so strongly to Copernicus that he repeated them in his own writings. As regards Pliny's recommendation about benefit received, "whatever this benefit," the young

Cracow author reported,Copernicus "admits that he received all of it from our university" of Cracow.

This statement has sometimes been understood to mean that Copernicus developed his own astronomy in Cracow, either by contact with the professors or with the books and manuscripts there. The two most distinctive features of Copernicus' astronomy concern the earth and the sun. Before Copernicus, the earth was generally believed to be stationary. He proclaimed that it is a planet in motion. Before Copernicus, the sun was generally considered to be a planet performing a daily as well as an annual revolution. Copernicus maintained that the sun stands still. Neither of these innovative propositions concerning the earth and the sun was available to Copernicus in Cracow.

Albert of Brudzewo's Commentary on George Peurbach's New Theory of the Planets. In the winter semester of 1491-1492 a general introduction to astronomy was taught at the University of Cracow. A year later, in the winter semester of 1492-1493, the same professor offered a more advanced course on the planets. The *New Theory of the Planets* by George Peurbach (1423-1461) was printed about 1472. In 1482 the older medieval *Theory of the Planets* was discarded in Cracow, where it was replaced by a commentary on Peurbach's *New Theory.* This commentator was Albert of Brudzewo, probably the most popular professor in the University of Cracow at that time.

To explain why the planets sometimes look bigger and at other times smaller to us on earth, Peurbach's *New Theory* depicted them as being carried eccentrically around the earth. Brudzewo's commentary maintained:

> No mortal man knows whether these eccentrics really exist in the spheres of the planets. Some people claim that the eccentrics, like the epicyles, are made manifest by the revelation of spirits. If we reject this claim, then the eccentrics are devised solely by the imagination of the astronomers.[46]

Richard of Wallingford. From the most original English mathematical astronomer of the later Middle Ages, Brudzewo quoted a like-minded passage:

> In the heavens there are no such eccentrics or epicycles as are devised by the astronomical imagination for its own use. No educated person could regard them as probable. Without such imaginative astronomical constructions, however, no systematic science of the motion of the stars can be established which would so pinpoint their positions at any moment as to be in accord with what we see.[47]

Immediately after this quotation, Brudzewo expressed his agreement:

> We should therefore be satisfied with this method since by means of it we obtain a perfect science of the stars in motion.[48]

Without being physically real, eccentrics and epicycles were indispensable intellectual constructs for the astronomer. For the geographer, the terrestrial equator is equally indispensable, although no such circle actually wraps itself around the earth.

Richard of Wallingford's *Albion.* Brudzewo called the author with whom he agreed "Albion." In so doing, he fell victim to a widespread confusion. Although printing had already been invented, Brudzewo had the work with which he agreed in the form, not of a printed book, but of a handwritten manuscript entitled *Tractatus Albionis*. This was understood by Brudzewo (and many others) to mean "A Treatise by Albion." At the end of a manuscript, the author or copyist often wrote *Explicit.* In this case the manuscript reads *Explicit Albion,* interpreted by Brudzewo to mean "Albion ends." The manuscript is divided into four parts, and Brudzewo quoted from Part I, Conclusion 10. In the prologue preceding Part I, Conclusion 1, the manuscript's opening statement is a description: "The Albion is a geometrical instrument."

For some reason this description was overlooked by Brudzewo and others. The author, Richard of Wallingford (1291 or 1292 - 1336), who states that he finished the work early in 1327, does not mention his name anywhere in the manuscript. Hence for a long time he was called Albion by those who did not know his true name.

Too late to be used by Brudzewo was a version of the manuscript explaining that the instrument "is called Albion, which is the same as 'all by one,' " a single device capable of performing all the desired operations.[49]

Copernicus' Attitude toward Eccentrics and Epicycles. In Cracow, Brudzewo's commentary on Peurbach was the standard work on planetary theory while Copernicus was a student there. Did he accept Brudzewo's view that eccentrics and epicycles are indispensable, though unreal? Later on, in writing his own astronomical works, Copernicus avoided making any categorical or explicit statement regarding this elusive topic. In general, it was his lifelong habit to keep quiet about any subject concerning which he was not sure. In this respect he was of one mind with a recent philosopher who said: "What we cannot talk about, we must keep quiet about."[50]

Copernicus' Disciple Rheticus. Toward the end of his life Copernicus attracted his one and only disciple, George Joachim Rheticus (1514-1574). Copernicus had long refused to publish his masterly work *On the Revolutions of the Heavenly Spheres (De revolutionibus orbium coelestium)* because it might involve him in controversy. But Rheticus realized its immense importance to mankind. He therefore rushed his own *First Report (Narratio prima)* on the Copernican astronomy into print. To his *First Report,* Rheticus added an appendix "In Praise of Prussia." For Toruń, where Copernicus was born; Frombork, where he served as canon; and Gdańsk, where Rheticus' *First Report* was printed in 1540, were three of the important cities in Prussia.

Copernicus' Conversation with Tiedemann Giese. Rheticus, who taught mathematics at the University of Wittenberg, obtained a leave of absence to visit Copernicus and master his astronomy at the source. After arriving in Frombork, Rheticus fell ill. To aid his recovery, he was invited to take a rest from his studies and spend several weeks, in the company of Copernicus, with Tiedemann Giese (1480-1550), an old friend of the astronomer.

Giese was largely responsible for overcoming Copernicus' reluctance to publish the *Revolutions,* as his masterpiece may be called for the sake of brevity. Copernicus and Giese had engaged in earnest conversation about the basic problems of astronomy, as Rheticus learned from friends familiar with the whole matter. Giese had warned Copernicus to expect no better treatment than Ptolemy had received from Ibn Rushd. As Giese put it,

> Ibn Rushd, who was in other respects a philosopher of the first rank, concluded that epicycles and eccentrics could not possibly exist in the realm of nature, and that Ptolemy did not know why the ancients had posited spiral motions. Ibn Rushd's final judgment is: "Ptolemy's astronomy is nothing as far as existence is concerned, but it is convenient for computing the nonexistent."[51]

Copernicus' Agreement with Giese. "Epicycles and eccentrics could not possibly exist in the realm of nature," said Ibn Rushd, "who was in other respects a philosopher of the first rank," according to Giese. Copernicus' friend thereby implied that epicycles and eccentrics could exist in the realm of nature. In Rheticus' account of the conversation between the two friends, Giese talks about the epicycles and eccentrics, whereas in his characteristic manner Copernicus says nothing about them. Nevertheless he agreed with Giese.

For the moon and the planets in the *Revolutions,* Copernicus used epicycles and eccentrics. He was, of course, aware of the equivalence between two alternative routes that could be projected for a heavenly body. It could be mounted on an epicycle whose center traveled around a deferent concentric with the center of the universe. Or the heavenly body could be mounted on the corresponding eccentric. At any given moment the heavenly body would be in the same place under either arrangement. "Hence, it is not easy to decide which of them exists in the heavens," Copernicus concluded, believing that one of them does. *(See Reading No. 8.)* He did not foresee what one of his greatest followers proved later: a planet's orbit is an ellipse. Copernicus still accepted the traditional dogma that a planet could move only in a circle. Only a circle could repeat the past, and the planets would roll on forever.

Copernicus' Disagreement with Ibn Rushd. For Ibn Rushd, Aristotle was not simply a philosopher, he was The Philosopher. In Aristotle's cosmos, every heavenly body had to move around the center of the universe. But every epicycle and every eccentric moved around a center different from the center of the universe. Therefore, there could be neither epicycles nor eccentrics in the cosmos inherited by Ibn Rushd from Aristotle.

When Copernicus was a student at the University of Cracow, the dominant philosophy was Aristotle's. But Copernicus did not share Ibn Rushd's attitude toward Aristotle. For example, Aristotle's theory of motion was unacceptable to Copernicus. Aristotle distinguished three types of motion: downward, toward the center of the universe, as in the case of earthy objects and water; upward, away from the center of the universe, as in the case of air and fire; and circular, around the center of the universe, as in the case of the heavenly bodies. The center of Aristotle's universe was the immovable earth.

For Copernicus, however, the earth was a planet revolving at a considerable distance from the center of the universe. But heavy objects on earth still continued to fall toward the earth's center, which was no longer the center of the universe. Such a heavy object, while falling downward, was being carried around circularly by the earth. Hence, the object was affected by two motions at the same time. This was the reason why Copernicus rejected Aristotle's theory of motion:

> Surely Aristotle's division of simple motion into three types, away from the middle, toward the middle, and around the middle, will be construed merely as a logical exercise. In like manner we distinguish line, point, and surface, even though one cannot exist without another, and none of them without body.[52] *(See Reading No. 9.)*

Because Copernicus deviated from Aristotle's theory of motion, he felt no compunction about admitting the non-Aristotelian epicycles and eccentrics to his astronomy.

The Nature of the Sphere in Copernicus' Astronomy. Copernicus entitled his masterpiece *De revolutionibus orbium coelestium, On the Revolutions of the Heavenly Spheres.* The title's third word *orbium* (of which *orbis* is the nominative singular) meant "spheres." But what kind of spheres? For instance, the heavenly bodies are spheres. A translator of Copernicus' *Revolutions* into German equated *orbium coelestium* with *Weltkörper*,[53] cosmic bodies, such as Venus and Mars. But Copernicus' *orbium* referred, not to such a sphere as Venus or Mars, but to a kind of sphere that had been banished from astronomy long before *orbium coelestium* was matched with *Weltkörper*. The spheres intended by Copernicus were invisible carriers of the visible planets. These are regarded today as moving through space without being attached to anything. But in the time of Copernicus (and long before him) a visible planet was thought to be attached to an invisible sphere *(orbis)* that transported the visible planet. These invisible spheres performed the revolutions mentioned by Copernicus in the title of his *Revolutions*.

Were Copernicus' Spheres Solid and Impenetrable? Copernicus' spheres, we were recently told, were "nonintersecting, rigid, material spheres, that is...solid spheres."[54] Nowhere in the *Revolutions* or in any other of his writings did Copernicus indicate, either explicitly or implicitly, that his spheres were solid, rigid, material, and nonintersecting. "The intersection of spheres is not permitted."[55] This is what we were recently told, but Copernicus never said it. The expressions *sphaera solida* and *orbis solidus*, never used by Copernicus, were employed by two later astronomers, who misattributed solid spheres to Copernicus.

The Attitude of Copernicus' Predecessors and Contemporaries toward Solid Spheres. Copernicus' "acceptance of solid spheres was as complete as that of any other astronomer before Tycho" Brahe (1546-1601). This complete acceptance, we were also recently told, included "the assumption, shared by all astronomers of Copernicus' time, that all planetary motions are controlled by the rotation of solid spheres." In other words, we were asked to believe that all astronomers before Copernicus' time and in Copernicus' time accepted "an assumption as universal and fundamental as solid spheres."[56]

Of all the astronomers before and in Copernicus' time, the only one put forward as having accepted solid spheres was Peurbach in the *New Theory of the Planets*. Not a word in Peurbach's *New Theory* indicates that his spheres were solid. That inference was drawn from his diagrams. But if his diagrams depicted non-solid spheres, how different would they look?

Brudzewo's commentary on Peurbach, the planetary textbook at the University of Cracow in Copernicus' time, dismissed the eccentrics as imaginary, not solid. When Copernicus moved to Italy to continue his studies, he encountered a book which said:

> Certain persons, and especially the Aristotelians...maintain that there are no eccentric, no epicyclic globes because, they say, the true and solid bodies of the planets cannot be transported by circles and lines that are drawn and lack body, so that a body cannot be coerced by an incorporeal

> body, nor are even epicycles to be called bodies lest there seem to be unoccupied [space] in heaven.[57]

The author of this Latin book did not indicate that he was translating Proclus' *Hypotyposis* from the Greek. The influence exerted by Proclus through the intervening centuries was immeasurable. His reference to the Aristotelians' rejection of eccentric and epicyclic globes foreshadowed Ibn Tufayl, Al-Bitruji, Ibn Rushd, Maimonides, Thomas Aquinas, Richard of Wallingford, Buridan, Henry of Hesse, Brudzewo, and how many others?

According to Ptolemy, a Heavenly Body Moved Without Interference by Any Other Heavenly Body. In the *Syntaxis* Ptolemy made a very sharp distinction between things on earth and those in the heavens. When things on earth move about, they encounter obstacles everywhere. In the heavens, on the other hand, there are no impediments. A heavenly body executes its motion without meeting any obstructions. No other heavenly body gets in its way.

Copernicus began to write the *Revolutions* at just about the time when Ptolemy's *Syntaxis* was printed for the first time. The copy annotated by Copernicus is still preserved. This earliest appearance of Ptolemy in print did not present the original Greek. Instead, it offered a translation into Latin that had been made in the twelfth century. In Book XIII, Chapter 2, Copernicus could read:

> There is no greater difference than that between things which are blocked on all sides, and things which are blocked neither by other things nor by themselves.[58]

Copernicus' Heavenly Spheres Are Interpenetrable. In the *Revolutions,* Book I, Chapter 10, Copernicus diagrammed his universe, which was finite. Its outermost boundary was formed by the "immovable sphere of the fixed stars," to which he assigned the number 1. Saturn, Jupiter, and Mars were given numbers 2, 3, 4, in that order. "The earth, together with the moon," received number 5.[59]

When the *Revolutions* was printed in Nuremberg under Rheticus' supervision, however, Copernicus' manuscript diagram was slightly changed. The moon's path was drawn encircling the earth. The "annual revolution" was ascribed to "the earth, together with the lunar sphere" *(cum orbe lunari).* Copernicus' accompanying text called the lunar sphere the earth's epicycle *(cum orbe lunari tanquam epicyclo).* The Nuremberg diagram depicted Copernicus' lunar sphere penetrating the region of his earth's sphere, and his earth penetrating the region of his moon's sphere. Two of Copernicus' heavenly bodies, earth and moon, have mutually interpenetrable spheres. So presumably have all his other heavenly bodies. Where, then, are Copernicus' "nonintersecting, rigid, material,...solid spheres"?

Notes

See p. 54 for Notes 42–45.

46. *Commentariolum super Theoricas novas planetarum Georgii Purbachii in studio generali Cracoviensi per Mag. Albertum de Brudzewo diligenter corrogatum A.d. 1482,* ed. Ludwik Antoni Birkenmajer (Cracow, 1900), p. 26/11-15.

47. John David North, *Richard of Wallingford* (Oxford, 1976), I, 278. The differences between North's text and Brudzewo's quotation are due to the diffusion of Richard's treatise in different versions (North, II, 127).

48. Brudzewo, *Commentariolum,* ed. L.A.Birkenmajer, p. 27/10-12.

49. North, *Wallingford,* II, 138.

50. Ludwig Wittenstein, *Tractatus logico-philosophicus* (London, 1961), p. 150.

51. Edward Rosen, *Three Copernican Treatises,* 3rd ed. (New York, 1971), pp. 194-195.

52. Nicholas Copernicus, *On the Revolutions,* translation and commentary by Edward Rosen (Baltimore, 1978), Book I, Chapter 8, p. 17/24-27.

53. Carl Ludolf Menzzer, *Nicolaus Coppernicus...Ueber die Kreisbewegungen der Weltkörper* (Thorn, 1879; reprint, Leipzig, 1939).

54. *Archives internationales d'histoire des sciences,* 1976, *26:* 110.

55. *Proceedings of the American Philosophical Society,* 1973, *117:* 467.

56. *Archives internationales d'histoire des sciences,* 1976, *26:* 114, 120.

57. Giorgio Valla, *De expetendis et fugiendis rebus* (Venice, 1501), sig. gg 7/20up - 15up.

58. *Almagestum Cl. Ptolemei* (Venice, 1515), fol. 142/8up - 6up.

59. *Nicholas Copernicus Complete Works,* I: *The Manuscript of Nicholas Copernicus' On the Revolutions, Facsimile* (London/Warsaw, 1972), fol. 9v.

CHAPTER 5

Copernicus in Italy

After leaving the University of Cracow, Copernicus did not proceed directly to Italy in pursuit of an advanced degree. Instead, he went to Lidzbark ("Heilsberg" in German), the episcopal administrative center for the diocese of Varmia. In his uncle's spacious palace in Lidzbark, a legal document executed for the bishop on 22 February 1496 was witnessed by Copernicus. He was described as a "cleric of the diocese of Chełmno" ("Kulm" in German), although he had already been named a canon of Varmia. He did not use that title then, because his seat was being contested.

Copernicus at the University of Bologna. In the winter semester of 1496-1497, classes at the University of Bologna began on 19 October. Following in the footsteps of his uncle, to whom that university had awarded a doctoral degree in canon law, Copernicus enrolled as a student of the same subject. He joined the Society of German Students of Law. This "nation," as it was called, consisted of students "whose native language is German, even though their home is elsewhere."

About a year after his arrival in Bologna, Copernicus received the welcome news that all opposition to his confirmation as a canon of Varmia had been overcome. Instead of undertaking the long, tedious, and expensive journey home to accept his appointment in person, he hurried to the office of a local notary. There he authorized two proxies back home

> in his name and on his behalf to receive, accept, and opt any and all freeholds and estates, and whatever property, movable and immovable, rights, actions, income, and benefits are due to him from any canonries still vacant. *(Studi e memorie per la storia dell' università di Bologna,* 1920, *5:* 232-233).

Thus did Nicholas Copernicus take possession of the Varmian canonry which he retained throughout his life.

Copernicus Was Not a Priest. Although Copernicus was a canon for many decades, he steadfastly refused to become a priest. Even when threatened by his bishop with the loss of his emoluments unless he proceeded promptly to take higher orders and enter the priesthood, Copernicus declined to do so. His manifest avoidance of the priestly office vexed an archivist, who recently falsified the Bolognese proxy document. Using the traditional formula, the notary had indicated that Copernicus appeared in person before him *(personaliter constitutus)*. But the archivist, who had discovered this notarial document, when he published it, deliberately replaced *personaliter* by *presbiter*, thereby pretending that Copernicus had been made a priest.[60]

Copernicus' Astronomical Activity in Bologna. Copernicus matriculated at the University of Bologna as a student of law, both canon and civil. "Once matriculation had occurred, the student was free to attend courses, and the professors could not exclude him." Copernicus' first love was astronomy. For pursuing advanced work in that science, he had acquired the requisite foundations at the University of Cracow. In Bologna he had the great good fortune to encounter a highly competent astronomer, who was ingenious enough to conceive new ideas and courageous enough to publish them, even when they contradicted established authority. That original thinker's name was Domenico Maria Novara or da Novara (1454-1504). As Copernicus told Rheticus some forty years later, "he was not so much the pupil as the assistant and witness of observations of the learned Domenico Maria"[61]Novara. For a while Copernicus lived in Novara's home. Like many another poorly paid professor of that period, Novara welcomed students as paying guests in his household. As such a boarder, Copernicus found it less inconvenient to assist Novara in his observations at night.

Domenico Maria Novara and Ptolemy's Geography. In establishing close contact with Novara, Copernicus met, perhaps for the first time in his life, a mind that dared to challenge the authority of the most eminent ancient writer in his chosen fields of study. In a 1489 publication that attracted considerable attention, Novara maintained that the latitudes of Mediterranean cities were then 1° 10' greater than those recorded in Ptolemy's *Geography*. From this systematic increase in latitude, Novara inferred a slow progressive alteration in the direction of the earth's axis.

Novara thereby publicly and explicitly doubted the validity of Ptolemy's authority. If the earth's axis changed its direction, then the earth was not as absolutely destitute of motion as Ptolemy had insisted. Novara's forthright attack on the abiding correctness of Ptolemy could hardly have failed to make a deep impression on his house guest, pupil, assistant, and witness of observations, "who knew his theories thoroughly."

Domenico Maria Novara Was Not a Neoplatonist. Domenico Maria wrote his surname as "Novara" or "da Novara." He indicated thereby that Novara, a city in northwestern Italy, had been the home of his ancestors. One of them, however, had been invited to move eastward across the top of the Italian boot to Ferrara, where Domenico Maria was born. In his publications he usually styled himself "Domenico Maria da Novara of Ferrara (Ferrariensis)."

After centuries when Aristotelianism was the dominant philosophy in the Latin West, Platonism was revived. Marsilio Ficino (1433-1499) founded the Platonic Academy in Florence, where it was sponsored by the ruling Medici family. Ficino wrote a letter to a thinker in Ferrara, whom he addressed as Ferrariensis and called his "fellow-philosopher."[62] This Ferrarese Neoplatonic philosopher, whose name was Francesco Marescalchi (d. 1482), has been mistakenly identified with Domenico Maria Novara Ferrariensis (d. 1504):

> Copernicus' friend and teacher at Bologna, Domenico Maria de Novara, was a close associate of the Florentine Neoplatonists.[63]

Domenico Maria Novara, Copernicus' teacher at Bologna, was an astronomer and not a Neoplatonist.

Copernicus Was Not a Neoplatonist. The author who misidentified Novara as a Neoplatonist went on to say:

> When Novara's pupil Copernicus complained that the Ptolemaic astronomers "seem to violate the first principle of uniformity in motion" and that they have been unable "to deduce the principal thing - namely the shape of the Universe and the unchangeable symmetry of its parts," he was participating in the same Neoplatonic tradition.

This author does not name a single Neoplatonist who voiced these objections to the Ptolemaic astronomy. It violated the first principle of uniform motion by introducing the equant. This device let a point revolve on a circle's circumference at a speed measured from outside the circle's center. The revolving point's distance from the circle's center was uniform, but its speed was nonuniform as seen from the circle's center. This combination of uniform distance and nonuniform speed violated the principle of uniform motion. The equant's violation of the principle of uniform motion was Copernicus' chief reason for breaking away from the Ptolemaic astronomy. He did so as an innovative astronomer, not as a traditional Neoplatonist philosopher.

Copernicus' Eulogy of the Sun. Before Copernicus, the cycle of the four seasons was generally explained by an annual revolution of the sun around the earth, which was regarded as stationary in the center of the universe. Copernicus interchanged the places of the earth and sun. His sun was in (or near) the center of the universe. He felt very strongly that as the major source of the heat and light on the earth, the sun did not belong where it had been put by the Ptolemaists, in the fourth slot among the bodies running around the earth. In the *Revolutions,* Book I, Chapter 10, after listing the six (then known) planets and their periods, Copernicus exclaims:

> At rest, however, in the middle of everything is the sun. For in this most beautiful temple [the universe], who would place this lamp in another or better position than that from which it can light up the whole thing at the same time? For, the sun is not inappropriately called by some people the lantern of the universe, its mind by others, and its ruler by still others.

> [Hermes] the Thrice Greatest labels it a visible god, and Sophocles' Electra, the all-seeing.[64]

For this impressive eulogy of the sun, we are told that Copernicus' "authorities are immediately Neoplatonic."[65] He names only two. Sophocles, the famous Athenian dramatist, wrote all his plays before Plato composed his earliest dialog. Hence Sophocles is pre-Platonic rather than "immediately Neoplatonic." He calls the sun all-seeing, not in his *Electra*, but in his *Oedipus at Colonus*, line 869.

Copernicus knew even less about his only other "immediately Neoplatonic" authority. This weird collection of spurious theological treatises disguised its true nature by claiming to have been written by the Greek divinity Hermes, whom Copernicus does not so much as mention. He knew Greek well enough to realize that Thrice Greatest should be Trismegistus. Yet he called the (supposed) author "Trimegistus." Nowhere in that sprawling body of writings attributed to Hermes Trismegistus is the sun called a visible god, as Copernicus asserts that it was. Evidently he had no first-hand acquaintance with the Hermetic collection. Pre-Platonic Sophocles and misquoted "Trimegistus" are Copernicus' "immediately Neoplatonic" authorities.

Novara and Ptolemaic Planetary Theory.

> Novara himself was among the first to criticize the Ptolemaic planetary theory on Neoplatonist grounds, believing that no system so complex and cumbersome could represent the true mathematical order of nature,

we were recently told.

> Domenico da Novara held that no system so cumbersome and inaccurate as the Ptolemaic had become could possibly be true of nature.[66]

Not a word, not a line, of Novara's writings was adduced to justify this portrayal of him as an opponent of the Ptolemaic planetary theory.

On the other hand, a student who visited the University of Bologna reported that Novara was "lecturing on Ptolemy's Almagest,"[67] as his *Syntaxis* was then called. Such lectures expounded the book under discussion, they did not condemn it.

Ficino Was Not a Copernican before Copernicus.

> Ficino wrote...that the sun was created first and in the center of the heavens. Surely no lesser position in space or in time could be compatible with the sun's dignity and creative function. But the position was not compatible with Ptolemaic astronomy, and the resulting difficulties for Neoplatonism may have helped Copernicus to conceive a new system, constructed about a central sun.[68]

Was Copernicus helped by Ficino, as suggested above? Copernicus conceived a new astronomical system, constructed about a sun permanently motionless in the center of the universe. In 1493, three years before Copernicus' arrival in Italy, Marsilio Ficino published a short essay *On the Sun.* He entitled Chapter X "The Sun Was Created First, and in the Center of the Heavens."[69] Ficino's sun was in the center of the heavens, not permanently like Copernicus' sun, but only at the instant of creation. In those days the account of creation in the Hebrew Bible was accepted as literally true. Few people were aware that there are two, rather different, Biblical accounts of creation.

When Ficino said that at the time of creation the sun was in the center of the heavens, he was referring to the pivotal zodiacal sign, the Ram. What happens to Ficino's sun after creation? It does not remain in the sign of the Ram, but makes a yearly trip through the twelve zodiacal signs. Ficino talks about the "entrance of the sun in [the signs of] the Crab, the Balance, and the Goat."

> The sun marks the four seasons of the year by [those] four changing signs....The motion of the sun as the first and principal planet is the simplest, since it does not deviate from the middle of the zodiac [that is, it has no celestial latitude], as the other planets do, nor does it retrogress,[70]

as the other planets do. Ficino's sun, like Ptolemy's, revolves around the center of the heavens. It does not occupy that center. On the contrary, it is a planet, revolving in a year, whose four seasons it marks off as it enters the Ram, Crab, Balance, and Goat. All these properties of Ficino's sun are thoroughly compatible with Ptolemaic astronomy, and present no difficulties whatever for Neoplatonism. They did not in any way help "Copernicus to conceive a new system constructed about a central sun." He did not feel the slightest hesitation in naming his predecessors whose earth revolved about a stationary central sun. Ficino was not one of these predecessors. In fact, there is no evidence that Copernicus ever heard of the little essay *On the Sun,* which Ficino composed in the autumn of 1491 and was printed on 31 January 1493.[71]

As a Student, Copernicus Ran Out of Funds. Bologna was a university town. Students had to obtain lodgings and meals from the townspeople. Rents were controlled by mixed commissions, composed of students and home owners. If prices rose too high, to prevent them from skyrocketing, the student organization, which controlled the university, as a last resort could threaten to move the institution elsewhere. At that time the University of Bologna owned no buildings of its own. Hence it could pack up and go to any other town offering better terms. The threat of such a secession deterred the business people of Bologna from squeezing the students too hard. But in this struggle between town and gown, for the most part the townspeople had the upper hand. The students managed as well as they could. From time to time, however, they ran out of funds. This is what once happened to Copernicus and his older brother Andrew.

Andrew Copernicus. Nicholas Copernicus the astronomer was the youngest of four children. He had two older sisters, and an older brother, Andrew. When Nicholas entered the University of Cracow in the winter semester of 1491-1492, Andrew matriculated at the same time. Nicholas paid the full fee, but Andrew only part of it. In itself, this difference may seem to be merely a minor detail. But in a certain sense it foreshadows the sharp contrast between the careers of the two brothers. Thus, Nicholas entered the University of Bologna in the winter semester of 1496-1497, but his older brother Andrew only two years later. Their uncle succeeded in procuring for Nicholas a Varmia canonry that fell vacant in 1495, whereas Andrew had to wait more than three years for a vacancy that opened up on 23 December 1498. A further divergence between the two brothers shows up when 100 ducats were borrowed from a bank for a period of four months in 1499 to tide them over a financial emergency. *(See Reading No. 10.)*

MAGISTER **Copernicus.** Before this loan was arranged, Andrew Copernicus thought of abandoning his studies in Bologna to look for a job in Rome. His younger brother Nicholas had no such thought. On 18 June 1499 he appeared in a notary's office in Bologna to act as a witness. In this document he was given the title of *magister* (master). This does not signify that he had received the academic degree "master of arts," as has sometimes been thought. In fact, he left the University of Bologna, as he had already left the University of Cracow, without obtaining any degree.

The University of Bologna resorted to the practice of employing a temporary teacher *(magister).* He taught recitation sections (they were called "repetitions") after a professor with contractual tenure had delivered a lecture. Such a *magister* resembles our own graduate students who are candidates for an advanced degree while teaching undergraduates at the same time. Copernicus may have taught elementary mathematics to inadequately prepared students of the liberal arts, in an educational setting not unlike our remedial courses. With this income, he did not feel as hard pressed as his brother Andrew to leave Bologna and hunt for paid work in Rome. Andrew was not as good a student as Nicholas, and was never appointed a *magister.*

Copernicus in Rome. Nicholas Copernicus acquired two astronomical works which he had a bookbinder combine in a single volume: a copy of the Venice 1492 edition of the *Alfonsine Tables,* and of a treatise by Regiomontanus (Augsburg, 1490).[72] On the title page of the *Alfonsine Tables* he wrote his name as owner. Behind Regiomontanus' treatise he had the bookbinder include a sheaf of sixteen sheets. On the verso of the sixteenth sheet two observations made in Bologna are recorded. They both concern conjunctions of the moon with Saturn in 1500, in the early morning of 9 January and 4 March. The composite volume is still preserved in the library of the University of Uppsala in Sweden. Handwriting experts are not sure whether these two observations were written down by Copernicus himself or by one of his associates. In any case, he was in Bologna in the early months of 1500. Classes at the University of Bologna ended on 6 September. Thereafter he went to Rome in the jubilee year proclaimed by the pope.

Copernicus as *Professor Mathematum*. While Copernicus was in Rome, as he told Rheticus nearly forty years later,

> when he was twenty-seven years old, more or less, about the year 1500, he lectured on mathematics in Rome before a large audience of students and a throng of great men and experts in this branch of knowledge.[73]

Rheticus included this statement in his *First Report* when he listed Rome as one of the three places where Copernicus made astronomical observations, the other two being Bologna and Frombork. Writing in Latin, Rheticus applied the expression *professor mathematum* to Copernicus. According to some biographers of Copernicus and historians of astronomy, Rheticus' expression means that Copernicus was a professor of mathematics or astronomy in Rome. A painstaking examination of the records of the University of Rome yielded no evidence that Copernicus had ever been appointed a professor at the *Sapienza,* as the University of Rome was affectionately called. Rheticus' *professor mathematum* simply indicated that Copernicus lectured publicly in Rome.

In 1539, when Copernicus told Rheticus about Rome, he could not remember the exact year when he lectured, nor his precise age at that time. He recalled being "twenty-seven years old, more or less, about the year 1500." He celebrated his twenty-seventh birthday on 19 February 1500, before he left Bologna for Rome. Hence he was more, not less, than twenty-seven when he addressed his Roman audience.

Did Copernicus Expound Geokineticism in Rome? What did he say to that distinguished audience? It has been thought that he expounded his new geokinetic and heliocentric astronomy. Had he done so, some members of that "throng of great men and experts in this branch of knowledge" would surely have reported his novel views, with approval or disapproval, in their publications and correspondence. Copernicus' public appearance in Rome produced not so much as a single echo from those highly articulate auditors, avid for every intellectual novelty. Their silence signifies that Copernicus spoke or wrote about his new astronomy only after he left Rome.

He observed a lunar eclipse in the early morning of 6 November 1500, while he was in Rome.[74] He may have stayed there during the opening months of 1501. That is why he told Rheticus that he talked in Rome "about the year 1500." In any case, he was back in Frombork, together with his brother Andrew, on 27 July 1501.

Copernicus Applies for a Leave of Absence to Study Medicine. On 27 July 1501 the two brothers appeared before a meeting of the Varmia Chapter, to which they both belonged as canons. The Chapter's statutes required any canon without a bachelor's or master's degree to study at a recognized university at least three years. While absent on leave from the Chapter for educational reasons, the student canon was entitled to receive his regular share of the income distributed to the sixteen canons every year. If the student canon performed well at the university, and wanted to extend his leave of absence, the Chapter would interpose no objection. *(See Reading No. 11.)*

Nicholas Copernicus, having already spent three years studying law with the Chapter's consent, at the meeting of the Chapter on 27 July 1501 asked for an extension of two years. None of his fellow canons had received any training in medicine. Nor were there any physicians in the vicinity who could be summoned in case of sickness or an emergency. As the canons grew older, they were prone to illness, with nobody qualified to heal them. Here was an opportunity for Nicholas to ingratiate himself with his fellow canons. He would study medicine, and come back prepared to take care of their ailments. He asked them for an extension of only two years. Since a medical school required three years of attendance for a degree, Nicholas Copernicus did not plan to become an M.D.

The situation of his brother Andrew was entirely different. He too had been a student of law. But he had not obtained the Chapter's prior consent. For on 27 July 1501 he asked the Chapter "for approval to begin his period of study, and continue it." Whereas the Chapter enthusiastically endorsed Nicholas' plan to undergo medical training, it conceded that "Andrew also seemed qualified to engage in studies." *(See Reading No. 12.)*

Copernicus' Second Trip to Italy. With the Chapter's consent, Nicholas Copernicus returned to Italy. This time he did not go to the University of Bologna. Instead, he chose the University of Padua, then the most prestigious center of medical studies in the Latin West.

While Copernicus was in Padua, he received word that he had been appointed the teacher (or scholaster, as this official was then called) in the Church of the Holy Cross in Wrocław (formerly Breslau). To take possession of this scholastry, Copernicus did not have to go to Wrocław in person. Instead, he went to the office of a notary in Padua on 10 January 1503. There he designated two Wrocław ecclesiastics as his proxies to act on his behalf. *(See Reading No. 13.)*

Copernicus' Scholastry in Wrocław. Just as Copernicus did not have to go to Wrocław to take possession of his scholastry, so he never went there in the exercise of this office. If any students were actually taught in the Church of the Holy Cross in Wrocław during Copernicus' scholastry, he personally did none of the teaching. That instruction would be taken care of by a substitute (or "vicar," as he was called), chosen by Copernicus, and paid by him. The difference between the vicar's salary and the income received by the scholastry accrued to Copernicus. If there were no pupils and no vicar, the entire income allotted to the scholastry would be retained by Copernicus. He held on to this office for thirty-five years before resigning from it toward the end of 1537 or the beginning of 1538. During all that time he never set foot in Wrocław. Members of his family, however, and of his uncle's family had held various offices in the Church of the Holy Cross in Wrocław. Moreover, the bishop of Varmia had the right of advowson - the right to present a nominee for an ecclesiastical vacancy - in the Church of the Holy Cross in Wrocław. At the time of Copernicus' appointment to the scholastry, the bishop of Varmia was his uncle.

Copernicus, Doctor of Laws. By studying medicine two years in the University of Padua, Copernicus used up the extended leave granted him by the Varmia Chapter on 27 July 1501. He had not qualified for the doctoral degree in medicine, which required a third year of study. But he had spent the number of years demanded for a doctoral degree in law at the University of Bologna. Nevertheless he did not return to Bologna to receive his doctoral diploma. Nor did he arrange to have the diploma awarded to him by the University of Padua. Instead, he switched to the University of Ferrara.

There, on 31 May 1503, he received the doctorate in canon law. *(See Reading No. 14.)* It may seem strange today that, after studying in Bologna and Padua, he took his diploma in Ferrara. The reason for his choice was the comparative cost of the festive ceremonies connected with the award of the doctoral diploma. Many of the students who flocked to Bologna and Padua came from families in high places. They could afford elaborate banquets to celebrate the successful termination of their studies, and expensive gifts to those who had helped them.

Copernicus' situation, however, was quite different. He was only ten years old when his father died. Had he not been lucky enough to have a prosperous uncle who liked him and pushed him along, he might never have gone to the Universities of Cracow, Bologna, Padua, and Ferrara. Without these contacts, would he have reconstructed astronomy? If he had not done so, would anyone else have tackled the job, then or later?

In any case, Copernicus felt that the conspicuous expenditures customary at Bologna and Padua were not for him. He had not forgotten that in 1499 he and his brother had to borrow a hundred ducats just to get along in Bologna. Hence, in 1503 he chose a humbler, albeit quite satisfactory, university for his doctorate. With his Ferrara diploma in hand, he returned to Varmia, the chief source of his lifelong income.

Notes

60. Edward Rosen, "Copernicus Was Not a Priest," *Proceedings of the American Philosophical Society,* 1960, *104:* 650-656; "Copernicus' Alleged Priesthood," *Archiv für Reformationsgeschichte,* 1971, *62:* 91-92.

61. Rosen, *Three Copernican Treatises,* p. 111.

62. Marsilio Ficino, *Opera omnia* (Turin, 1959, 1962; reprint of Basel 1576 ed.), pp. 644, 738, 776; *The Letters of Marsilio Ficino* (London, 1975-1978), I, 125; II, 50; *Two Renaissance Book Hunters,* tr. Phyllis W. G. Gordan (New York/London, 1974), pp. 215-217, 219.

63. Thomas S. Kuhn, *The Copernican Revolution* (Cambridge, MA: Harvard University Press, 1957; frequently reissued), p. 128.

64. Nicholas Copernicus, *On the Revolutions,* p. 22/3-8.

65. Kuhn, *Copernican Revolution,* p. 130; Edward Rosen, "Was Copernicus a Neoplatonist?", *Journal of the History of Ideas* , 1983, *44:* 667-669.

66. *Ibid.*, p. 128; Kuhn, *The Structure of Scientific Revolutions,* 2nd ed., enlarged (Chicago, 1970), p. 69.

67. *Der Briefwechsel des Konrad Celtis,* ed. Hans Rupprich, p. 438/51-53 (Munich, 1934; Veröffentlichungen der Kommission zur Erforschung der Geschichte der Reformation und Gegenreformation, Humanistenbriefe, III).

68. Kuhn, *Copernican Revolution,* p. 130.

69. Ficino, *Opera omnia,* p. 971.

70. *Ibid.*, pp. 966-967.

71. Paul Oskar Kristeller, *Supplementum Ficinianum* (Florence, 1937), I, cxii.

72. Paweł Czartoryski, "The Library of Copernicus," p. 366, #2, in *Science and History: Studies in Honor of Edward Rosen* (Wrocław, 1978; Studia Copernicana, XVI, ed. Erna Hilfstein *et al.).*

73. Rosen, *Three Copernican Treatises,* p. 111.

74. Copernicus, *On the Revolutions,* Book IV, Chapter 14, p. 200/31-32.

CHAPTER 6

Copernicus' Return to Varmia

Copernicus complied with the statutes of the Varmia Chapt
by returning from Italy with a doctoral degree in law. We have
direct trace of his movements during the seven months following h
receipt of the Ferrara diploma on 31 May 1503. But on the day cor
memorating Jesus' Circumcision, 1 January 1504, Copernicus ar
his uncle arrived in Malbork (Marienburg, in German) to attend
scheduled meeting of the Estates of Royal Prussia. King Alexand
of Poland, through his spokesman, summoned the members of t
Estates to swear their regular pledge of loyalty to him at the meeti
of the Polish Diet on 21 January 1504 in Piotrków Trybunalsk

Prussia and Poland. Originally the Prussians were a pagan pe
ple who spoke a non-Germanic language, related to Lithuanian, La
vian, and Estonian. In the German *Drang nach Osten* (push to t
east) the Prussians were subjugated and forcibly converted to Chr
tianity. In the course of time their language (Old Prussian) was su
pressed and replaced by German. Their rulers, the Teutonic Knigh
dealt harshly with all their subjects, whose reaction was the form
tion of the Prussian Union. This by itself did not have the streng
to throw off the yoke of the Teutonic Knights. To accomplish th
result, the Prussian Union sought the help of Poland. The negotiato
for the Prussian Union included Copernicus' father. In the Th
teen Years' War (1454-1466) the Teutonic Knights were defeated. Th
ceded West Prussia to Poland, which guaranteed limited rights
self-government to what became Royal Prussia in 1466. Nearly fo
ty years later, in the exercise of those rights the Estates of Roy
Prussia, meeting in Elbląg (Elbing, in German) on 18 January 150
asked King Alexander to receive the oath of loyalty while visiti
Royal Prussia. In due course, on 18 May, the king arrived in Elbl
to accept the pledge of allegiance from his Prussian subjects, i
cluding Copernicus and his uncle.

Nicholas Copernicus and Lucas Watzenrode. It seems clear th
after his return to Varmia from Italy in 1503, Nicholas Copernic
resided in the episcopal palace of his uncle, Bishop Lucas Watzenro
of Varmia, in Lidzbark (Heilsberg, in German), rather than with h
fellow canons in the cathedral town of Frombork. On 7 Janua
1507 the Varmia Chapter voted to give Copernicus 15 marks a yea
over and above his regular income as a canon, as long as he us
his medical knowledge to take care of his uncle's health. *(See Readi
No. 15.)* Under his uncle's guidance Copernicus petitioned the papa
to permit him to hold additional church benefices. On 29 Novemb
1508 his petition was approved by the pope, who permitted him
accept two additional church benefices, including those for which

was not officially qualified. In seeking and obtaining this papal permission Copernicus was treading in the path pursued by his uncle, who was eminently successful in accumulating a large number of sinecures - offices yielding an income without requiring any related work. The predictable outcome of this imitation of his uncle's grasping and go-getting behavior would be the eventual elevation of Nicholas Copernicus to the bishopric of Varmia, perhaps as his uncle's immediate successor.

Copernicus, Canon of Varmia. Copernicus never took advantage of the permission granted to him by the pope on 29 November 1508 to add to his Varmia canonry and Wrocław scholastry. He also forewent the bonus of fifteen marks voted him by the Chapter on 7 January 1507, when in the latter part of 1510 he decided to leave his uncle's palace in Lidzbark and live the ordinary canon's life in Frombork. At the annual meeting of the Varmia Chapter on 11 November 1510, Copernicus was elected chancellor of the Chapter. Together with another canon, he was named inspector *(visitator)* of the Chapter's holdings. In this capacity, on 1 January 1511, he and his fellow inspector visited the Chapter's southern stronghold of Olsztyn (Allenstein, in German). There they approved the accounts of the previous year, and returned to Frombork, bringing with them a sizable sum of money for the Chapter's treasury. *(See Reading No. 16.)*

Copernicus, Wealthy Ecclesiastic or Dissident Astronomer? Why did Copernicus forsake his uncle the bishop, and take up his duties as a canon? Could he have combined the pursuit of ecclesiastical preferment with the development of his intellect? Would not unremitting attention to the intricacies of church politics have interfered with his vision of a reconstructed astronomy?

Alternatively, we know of no personal quarrel between uncle and nephew that would account for Copernicus' decision to leave Lidzbark and live in Frombork. True, the life of a canon in Frombork was far from being a sinecure. But despite all its problems, at least it was free from the constant hustle and bustle of the episcopal palace in Lidzbark. It gave Copernicus access to a tower - not an ivory tower, to be sure, but an observatory tower. It granted him the opportunity to construct his own observational instruments. It offered him undisturbed time to think about what was right and what was wrong in current astronomical theory.

Copernicus' Private Life. Copernicus' canonry furnished him income enough to sustain the life of the mind, without involving him in an incessant chase after greater and greater wealth. After all, canon law forbade him to marry. Hence, he had no wife and legitimate children to support. Unlike his uncle and a later bishop of Varmia, he had no illegitimate children to support, either.

His principal expense was his housekeeper. A Varmia canon had his own private quarters. He did not clean them himself, nor did he cook his own meals. For these purposes and related household tasks, he engaged a live-in housekeeper. Naturally, skeptical tongues wagged about what might be going on between an unmarried canon and his housekeeper. For this reason a female relative was preferred as less likely to give rise to a scandal.

The Woman who Stayed Overnight. A woman (not yet identified) worked as Copernicus' housekeeper. Of her own free will she left his employment to get married. Her decision to do that was not influenced by any action on the part of Copernicus. But her marriage was never consummated, because her husband turned out to be impotent. Even though Copernicus tried to persuade the couple to remain respectable spouses, they went ahead with two hearings before officials of the Chapter, and separated.

The woman later went to work for a matron in Elbląg. Together they visited the fair in Koenigsberg (now Kaliningrad in the Soviet Union). On their way back to Elbląg, they had to spend the night in Frombork, where construction of a tavern or new hotel was being financed. As a small return for his former housekeeper's faithful services, Copernicus permitted the two women to stay in his home.

Gossip being what it is, this matter was reported to the bishop in Lidzbark. He admonished Copernicus. But in order to conceal his reaction from his secretary, he wrote out the letter with his own hand. On 27 July 1531 Copernicus promised the bishop that in the future nobody would have any proper reason to suspect him of misconduct. *(See Reading No. 17.)*

Copernicus Is Advised to Change his Housekeeper. Bishop Maurice Ferber of Varmia died on 1 July 1537, and was succeeded by John Dantiscus. The new bishop decided to tour his diocese in order to receive the oath of allegiance from its inhabitants. To accompany Dantiscus on his rounds, the Varmia Chapter delegated two representatives, Copernicus and a fellow canon, Felix Reich. They met Dantiscus in the episcopal palace early in August 1538. Under these circumstances Dantiscus spoke to Copernicus as an individual about the general problem of the canons' employment of housekeepers, advising Copernicus to make a change. It was not easy, however, for Copernicus to find a suitable female relative right away to replace his housekeeper. He hoped to be able to do so by the following Easter, 6 April 1539. Angered by Copernicus' delay, Dantiscus sent him a severe reprimand, insisting on earlier compliance. On 2 December 1538 Copernicus replied that he had no intention of stalling: he would settle the matter by Christmas, 1538. *(See Reading No. 18.)*

Felix Reich's Letter to Dantiscus about Copernicus. When Dantiscus reprimanded Copernicus with regard to his housekeeper, he informed Felix Reich about the step he had taken. He also sent Reich a letter which, in accordance with their previous secret agreement, Reich was to read to Copernicus by way of intensifying the pressure on him. When Reich saw Dantiscus' letter, however, he refused to go ahead. As a notary licensed by the pope and the emperor, Reich was more familiar than Dantiscus with legal terminology. Dantiscus' "insertion of certain little words" in his letter deterred Reich from reading it to Copernicus.

In any case Reich felt very uncomfortable about letting Copernicus know that he was aware of what was going on. He had accompanied Copernicus on Dantiscus' tour of the diocese. His personal attitude toward Copernicus was disclosed in the will which he drew up on 22 November 1538, about the time he entered into the secret agreement with Dantiscus to read the bishop's letter aloud to Copernicus.

Reich was a man of considerable wealth. He could easily have provided for all fifteen of his fellow canons. In fact, however, he chose only three, the second being Copernicus, to whom he bequeathed four gold coins. "I gave him as his personal property while I was still alive"[75] a copy of Dioscorides' *Materia medica,* according to a remark in the testament of Reich, who had been one of Copernicus' patients. On 1 November 1538 Reich had informed Dantiscus: "Through God's mercy and the efforts of the doctor, Nicholas [Copernicus], my loss of blood was stopped in time."[76]

In agreeing to cooperate with Dantiscus in the bishop's move against Copernicus, Reich was motivated by considerations of internal Chapter politics rather than by feelings of antagonism toward Copernicus. This mingling of attitudes is revealed in the letter which Reich sent to Dantiscus on 2 December 1538, the day on which the courier from Frombork to Lidzbark also carried Copernicus' reply to Dantiscus. *(See Reading No. 19.)*

Reich's Campaign against Copernicus and Two Other Canons. On 11 January 1539 Copernicus wrote to Bishop Dantiscus:

> I have now done what I should not or could not in any way have failed to do. I hope that what I have done in this matter quite accords with your Most Reverend Lordship's warnings.[77]

In this discreet manner Copernicus informed Dantiscus that he had dismissed his housekeeper.

On the same day, 11 January 1539, Reich also wrote to Dantiscus about prosecuting three canons, including Copernicus. In particular, Reich spontaneously offered to help the bishop proceed against the doctor who had saved his life, as Reich himself had informed Dantiscus a little over two months earlier.

Reich's health had deteriorated so far that he could scarcely do any actual writing himself. But he was still able to dictate a writ. He recommended that the bishop should issue an order in accordance with this writ. The order should be sent as quickly as possible in a sealed letter addressed individually to each of these three canons.

Reich had previously sent the bishop another writ, according to which Dantiscus could have the local priest under his authority warn the three women involved. Two of them had no husbands, by contrast with Copernicus' housekeeper, who was married. Hence, the letter to her contained material which should be omitted from the letters to the other two. *(See Reading No. 20.)*

Dantiscus and Reich Join Forces against Copernicus. Felix Reich's letter, dispatched from Frombork on 11 January 1539, was received by Dantiscus in Lidzbark on 15 January. Since Reich had recommended that the bishop should act promptly, the legal documents were drafted by Dantiscus and his staff in a great hurry. Then they were sent to Reich, an experienced notary, to be scrutinized. When he examined them, he found serious flaws. For example, one of the canons under attack was Copernicus' friend, Alexander Scultetus. The document, however, called him "Henry." This was merely a scribal error. More serious, on the other hand, was a legal blunder. The bishop had no authority to banish anyone beyond the boundaries of his diocese. Yet the women were banished to ten miles from Frombork. It, however, was closer than ten miles from the northwestern boundary of the bishopric. Hence, Reich returned the defective drafts to Dantiscus. *(See Reading No. 21.)*

Reich Withholds Dantiscus' Letter to the Varmia Chapter. Reich's letter of 23 January 1539 was received by Dantiscus on 27 January. On that same day a shipment of wine and beer that had been ordered by Dantiscus was delivered to Reich. This shipment was accompanied by a letter addressed to the Varmia Chaper. Since this letter was not addressed to Reich, he would not open it. He was afraid, however, that Dantiscus' unopened letter to the Chapter might have something to say about the proceedings against the three canons and their cooks. In that case, if Reich delivered the letter to the Chapter, an uproar might break out. At the moment only three canons happened to be in residence, the others being temporarily absent. So small a number of canons could not with propriety rule on so serious a matter as the bishop's proceedings against the three canons and their housekeepers. Hence, relying on his own judgment, Reich decided to withhold Dantiscus' letter to the Chapter. He begged the bishop not to be angry with him for delaying the delivery of this letter. Without great loss of time, Dantiscus' letter could be redirected to the Chapter later on. *(See Reading No. 22.)*

Reich's letter concerning the withholding of Dantiscus' letter was dated "with sick and trembling hand" on 27 January 1539. It was received by Dantiscus on 30 January. A month later, on 1 March 1539, Reich, that ungrateful, timorous, devious, and disloyal "friend" of Copernicus, died.

Anna Schilling. On 11 January 1539 Copernicus notified Bishop Dantiscus that he had dismissed his housekeeper, Anna Schilling. She owned her own house in Frombork, which she wanted to sell before going to Gdańsk (Danzig, in German), her birthplace. "She sent her belongings ahead to Gdańsk, but is still living by herself in Frombork," as was reported to Bishop Dantiscus by the provost of the Varmia Chapter on 23 March 1539. This provost, Paul Płotowski, was the first Pole to be admitted to the Varmia Chapter. *(See Reading No. 23.)*

Tiedemann Giese. Bishop Maurice Ferber of Varmia died on 1 July 1537. To fill the vacancy, the Varmia Chapter sent a list of all the eligible canons to the king of Poland. He in turn chose a panel of four canons acceptable to himself. The king's panel, drawn up on 4 September 1537, included Copernicus and his closest friend, Tiedemann Giese. However, the candidate most strongly backed by the king was John Dantiscus (1485-1548), who was then the bishop of Chełmno. If elected by the Chapter from the king's panel to the more remunerative bishopric of Varmia, Dantiscus would leave the Chełmno diocese without a bishop. That vacancy was promised to Giese, if he withdrew his candidacy in Varmia. He did so. By this prearranged deal Dantiscus became bishop of Varmia and Giese bishop of Chełmno.

In his report of 23 March 1539, Provost Płotowski wrote to Dantiscus that Alexander Scultetus, a Varmia canon under attack with regard to his housekeeper, had returned to Frombork from Chełmno "with a happy countenance." Płotowski thereby implied that Giese was supporting Scultetus against Dantiscus. The bishop of Varmia wrote at once to Giese, asking what had made Scultetus so happy in Chełmno. Giese's reply in effect called Płotowski a liar, trying to stir up trouble between the two bishops. *(See Reading No. 24.)*

Giese Is Copernicus' Patient. About a year after being appointed bishop of Chełmno, a Royal Prussian diocese lying not far to the west of Varmia, Giese contracted a fever while visiting an area in the northern part of his bishopric. His illness was reported by his chaplain to Dantiscus. Medications were prescribed for Giese by two physicians, one from Toruń (Thorn, in German), Copernicus' birthplace, and the other from Gdańsk. Copernicus was also summoned from Frombork. He arrived in Lubawa on 27 April 1539, after the doctor from Toruń had left. Copernicus' prognosis, in agreement with the doctor's from Gdańsk, was that better days were in store for Giese. *(See Reading No. 25.)*

Dantiscus Tries to Enlist Giese's Help against Copernicus. An (unidentified) informer told Bishop Dantiscus of Varmia that Copernicus was expected to visit his friend, Bishop Giese of Chełmno.

On 5 July 1539 Dantiscus wrote to Giese, expressing the highest regard for Copernicus, whom he dearly loved and admired for his accomplishments. Yet he "is said to let his mistress in frequently in secret assignations." (Copernicus was then more than sixty-six years old.) Dantiscus urged Giese to warn Copernicus to stop seeing her. However, Dantiscus wanted Copernicus to believe that the warning originated with Giese, acting on his own, without any prior suggestion from Dantiscus. The bishop of Varmia wanted the bishop of Chełmno also to caution Copernicus in the same way against allowing himself to be led astray by Alexander Scultetus who, according to Copernicus, "all by himself is outstanding in all respects among all" the canons of Varmia. Scultetus was suspected by Dantiscus of harboring sympathies for the Protestant Reformation. *(See Reading No. 26.)*

Giese Undertakes to Warn Copernicus, who Will Be More Receptive if he Knows that the Warning Emanates from Dantiscus. Giese answered Dantiscus very promptly, on 7 July 1539. The bishop of Chełmno will wholeheartedly do what the bishop of Varmia wants. But Giese is confident that Copernicus will be more deeply affected if he understands that Giese is acting at the suggestion of Dantiscus. The advice is offered sincerely with a view to Copernicus' reputation, against those who may have convinced him otherwise. *(See Reading No. 27.)*

Copernicus Rejects Dantiscus' Charges, as Relayed by Giese. Bishop Giese found an opportunity to confront Copernicus with Dantiscus' charges after the arrival in Frombork of Rheticus, a professor of astronomy on leave from the University of Wittenberg. Having heard something about Copernicus' new cosmology, Rheticus decided to talk to the old master in person. The religious situation was precarious. Bishop Dantiscus had banished all sympathizers with Lutheranism from the diocese of Varmia on 21 March 1539. On 14 May 1539 Rheticus reached Poznań (Posen, in German) on his way to Frombork from Wittenberg, the think tank of Lutheranism. Yet Rheticus the Lutheran was welcomed by the Roman Catholic canon Copernicus. The manuscript of the *Revolutions* written by Copernicus' own hand was studied intensively by Rheticus. When he became ill, he was invited, together with Copernicus, to Lubawa by Bishop Giese of Chełmno. In this way Giese was able to confront Copernicus with Dantiscus' accusations. Shocked by these unfounded and malicious rumors, Copernicus denied having admitted Anna Schilling to his home after her dismissal as his housekeeper. She had spoken to him once in passing as she was leaving Frombork for the fair in Koenigsberg. Giese sternly cautioned Copernicus against providing his detractors with even the appearance of misconduct. Giese ended his letter to Dantiscus by reminding that severe critic of sexual misbehavior that, as the notorious father of an illegitimate daughter, he too might not be beyond the reach of calumnious tongues. *(See Reading No. 28.)*

Giese's Report to Dantiscus Is Confirmed by the Administrator of the Chapter. Another member of the Varmia Chapter who reported to Bishop Dantiscus about Copernicus' attitude toward Anna Schilling was Achatius von der Trenck, the Administrator of the Chapter with his headquarters in Olsztyn in the southwestern part of the diocese. Trenck had promised to visit Giese in Lubawa, and did so while Copernicus was there. Aware of the Giese-Copernicus conversation concerning Anna Schilling, Trenck also discussed her with Copernicus. He assured Trenck that "he would never receive her in his house." After returning to Olsztyn, Trenck reported to Dantiscus about his meeting with Copernicus in Lubawa. *(See Reading No. 29.)*

Bishop Dantiscus Chides the Varmia Chapter for its Inactivity regarding Heretics and the Canons' Female Employees. Dantiscus acted very vigorously in expelling Lutherans and other heretics from his diocese of Varmia. He also initiated strenuous measures directed against the women employed as housekeepers by the Varmia canons. Ignoring those who were well behaved, he denounced the lot as a bunch of whores. In a set of instructions relayed through Provost Płotowski to the Varmia Chapter at its general meeting about 1 November 1540, Bishop Dantiscus expressed deep indignation at the Chapter's failure to keep step with him in his campaign against Protestants and those he called prostitutes. *(See Reading No. 30.)*

The Case of Alexander Scultetus' House. In addition to Copernicus, a second Varmia canon under attack with regard to his household was Alexander Scultetus. In his youth he had been employed in the Roman Curia, where he acquired some influential friends. Hence, when an official complaint against him was submitted in 1539 by the king of Poland to the pope, nothing happened. After waiting in vain for a long time, the king took matters in his own hands. He ordered Scultetus to appear on 24 May 1540 to answer charges of being married, having several children, and belonging to the heretical Sacramentarians, who denied the physical presence of Jesus' body and blood in the eucharistic wafer, which in their view was called "body and blood" only sacramentally or metaphorically. Knowing full well what awaited him in a royal court dominated by a king already convinced of his guilt, Scultetus vanished, leaving behind an empty house in Frombork.

These premises were promptly occupied by a Polish nobleman, Nicholas Płotowski, undoubtedly related to Paul Płotowski, provost of the Varmia Chapter. Nicholas Płotowski's occupancy was contested by a certain Henry Braun in a legal action heard by the Varmia Chapter. Płotowski claimed that as a subject of the crown, he was not answerable to the Varmia Chapter. He therefore asked that the case should be referred to the king. In response, Braun argued that the king had turned the case over to the local authorities, the Varmia Chapter being the proper tribunal.

Nicholas Płotowski Tries to Disqualify Copernicus as a Judge. The first hearing in the Płotowski-Braun lawsuit over Scultetus' house took place on 10 January 1540. At that time Płotowski contended that both Nicholas Copernicus and Leonard Nidderhoff should withdraw as judges from the case. He would prove at the proper time and place that they were under suspicion, along with Scultetus. If they did not disqualify themselves, he would use their presence on the bench as the basis for an appeal to a higher court.

The Varmia Chapter decided to postpone the hearing until 23 April 1540, when Płotowski could produce his objections to Copernicus and Nidderhoff. When the Chapter reconvened on that day, however, it excluded the litigants, limited the number of canons serving as judges to no more than three, and invoked the legal rule holding that an objection to an individual is invalid unless it is made in his presence. But Copernicus was then far away in Koenigsberg. As a physician of wide reputation, he had been summoned by the duke of Prussia to treat an important patient. The Varmia Chapter had given Copernicus a leave of absence to stay in Koenigsberg for the sake of his patient. Hence, on 23 April 1540 it recessed the case of the Scultetus house for two months. *(See Reading No. 31.)*

To plead his own case, Scultetus went to Rome. But back home his strongbox was opened by special royal commissioners. They found copies of the *Commentaries on Paul's Epistles to the Hebrews* and *to the Romans* by the eminent Swiss reformer Henry Bullinger, with marginal notes by Scultetus. He spent three years in jail. After being released, Scultetus published a *Chronology or Annals of Nearly All the Kings, Princes, and Potentates from the Creation of the World to the Year 1545* (Rome, 1546). In his list of famous men, Scultetus did not forget to include his old friend, "Nicholas Copernicus, canon of Frombork, astronomer and mathematician."

Copernicus' Serious Illness. A new canon, George Donner, was inducted into the Varmia Chapter on 12 April 1540 in the presence of Copernicus. By contrast with Copernicus, Donner did not gain admission to the Chapter as a young man. He was already well along in years, and being himself elderly, he became friendly with Copernicus, who was then approaching seventy. When the astronomer fell gravely ill in 1542, it was Donner who notified Giese. From Lubawa, on 8 December 1542 the bishop of Chełmno replied to Donner, expressing his deep concern over Copernicus' illness. Giese asked Donner to take good care of the sick old man, lest he should be deprived of brotherly help in his perilous condition. *(See Reading No. 32.)*

When Copernicus drew up his last will and testament (which has not been preserved), he named Donner as one of his four executors, along with Nidderhoff, a third canon, and the father of the youthful distant relative who had been selected by Copernicus to succeed him in the canonry. Copernicus' *Revolutions* was printed in faraway Nuremberg, and a copy reached him on the day he died, 24 May 1543. Donner sent a copy to the duke of Prussia, who on 28 July acknowledged receiving it. As an executor of Copernicus' will, Donner outlived the testator less than a year.

The Execution of Copernicus' Will. Nicholas Copernicus was the youngest of four children. We have already heard something about his older brother, Andrew. Of his two sisters, one married and raised a family, including a daughter Christina. To this niece, Nicholas Copernicus bequeathed some property in his will. Her husband, Caspar Stulpawitz, was employed as a military drummer by the duke of Prussia. As soon as Stulpawitz learned that his wife was mentioned as a beneficiary in Copernicus' will, he asked his employer to give Christina Stulpawitz a letter addressed, jointly and severally, to the four executors of Copernicus' will. The duke's letter, dated in Koenigsberg, the capital of the duchy of Prussia, on 29 June 1543, was taken to Frombork by Copernicus' niece. *(See Reading No. 33.)* What was bequeathed to her is not known, since Copernicus' will has not been preserved.

Copernicus' Dying Days. In the absence of a printing press in Frombork, Rheticus arranged to have Copernicus' *Revolutions* printed in Nuremberg, where he served as editor. By the time the typesetters finished Book IV, however, Rheticus had to leave Nuremberg for Leipzig, where he had just been appointed professor of astronomy in a reorganization of that university. In Nuremberg, Rheticus' successor as editor of the *Revolutions* surreptitiously interpolated an unsigned Address to the Reader in flat contradiction to the book itself. Copies of the *Revolutions*, so maltreated, were sent from Nuremberg to Rheticus in Leipzig. Outraged by what he saw, Rheticus transmitted two copies to Giese. At that time the bishop of Chełmno happened to be away, since he was attending the wedding of Poland's crown prince in Cracow. On his return to Lubawa, Giese became just as indignant as Rheticus. Drafting a letter of protest to the City Council of Nuremberg, Giese asked Rheticus to pursue the matter, since he was more familiar than Giese with the people in Nuremberg.

Giese wanted the *Revolutions* reissued without the fraudulent Address to the Reader and with two brief additions by Rheticus. For, like Copernicus, Rheticus saw no incompatibility between the Bible, properly understood, and the Copernican astronomy, properly understood. Rheticus had composed a short treatise on this subject as well as a brief biography of Copernicus. The astronomer was still alive when Rheticus had left Frombork in the autumn of 1541. In Giese's letter of 26 July 1543 to Rheticus, the bishop of Chełmno supplied the details of Copernicus' dying days. *(See Reading No. 34.)*

Giese's protest to the City Council of Nuremberg did not produce a revision of the *Revolutions.* Rheticus' argument for the compatibility of the Copernican astronomy with the Bible has not been preserved. The same fate befell his biography of Copernicus.

The Fate of Anna Schilling after Copernicus' Death. At a trial held in the court of Bishop Dantiscus, the housekeepers of the three canons under attack, including Copernicus, were banished from the diocese of Varmia. But Anna Schilling, Copernicus' banished housekeeper, owned a house in Frombork. Since she could no longer reside in the diocese, she wanted to sell the house. For this reason, from time to time, she returned to Frombork and stayed a few days. She was said to have found a buyer on 9 September 1543. On the following day the Varmia Chapter wrote to Bishop Dantiscus, asking him how to deal with her. She had been banished on account of Copernicus, while he was alive. Now he is dead. According to the legal maxim, when the cause ceases, the effect ceases. Is it legal to exclude her any longer? *(See Reading No. 35.)*

Bishop Dantiscus answered angrily and promptly. The decision was up to the Varmia Chapter. However, according to the bishop, she ensnared Copernicus. She can do it again, to anyone of you. Keep her out. *(See Reading No. 36.)*

Notes

75. Marian Biskup, "Testament kustosza warmińskiego Feliksa Reicha z lat 1538-1539," *Komunikaty Mazursko-Warmińskie*, 1972, p. 657/3-5.

76. L.A.Birkenmajer, *Mikołaj Kopernik*, p. 392, #7.

77. Drewnowski, p. 234/#9/3-5; Studia Copernicana, *18*.

CHAPTER 7

COPERNICUS, ECONOMIST

Copernicus achieved immortal fame as an astronomer. What he contributed to economics is less well known. During his lifetime and in the region where he was active, paper money was not used. Neither were checks. Business transactions were conducted by means of coins made of metal. The principal metal used for this purpose was silver. But pure silver, however much it was admired when made into jewelry or bowls, platters, boxes and the like, had certain drawbacks when minted as money.

Why Silver Coins Were Alloyed with Copper. If a coin was minted from pure silver and put into circulation, it might be melted down by someone who could get more for the molten silver than the coin would buy. Even in modern times, when the price of silver climbs high enough, there are people who melt the silver down and sell it at a higher price than the coin. If the government did not intervene promptly and effectively, the silver coinage would soon disappear from the market place. This crippling effect on daily business life is averted by alloying silver with a base metal. In Copernicus' time, the base metal used for this purpose was copper.

Many ordinary business transactions require the exchange of only a very small amount of silver. The minting of such tiny coins from pure silver would make them inconvenient and virtually unmanageable. Hence they were alloyed with copper to give them a convenient size.

The constant passing of silver coins from hand to hand tends to wear them down. They last longer when alloyed with copper.

These three reasons for alloying silver with copper were discussed by Copernicus in a paper he wrote. It was given the title *Essay on the Coinage of Money* by his fellow canon Felix Reich, who like Copernicus became a spokesman of the Varmia Chapter on the money question.

Why Precious Metals Were Coined. All over the earth and throughout the ages, people have attached the highest importance to acquiring and holding gold and silver as personal possessions. Before coinage was introduced, the precious metals were used to pay for important purchases. A businessman had his own set of weights and his own balance. On one of its pans, the metal was placed to equalize a standard weight on the other pan. However, transporting the weights and the balance, a delicate instrument, from place to place was bothersome. Moreover, some merchants were suspected of having two sets of weights, one slightly heavier and the other slightly lighter, than the accepted standard. In addition how could the ordinary person detect impurities in the gold or silver lying on the balance's pan?

Governments undertook to get rid of these problems by issuing minted coins. These bore the symbol of the issuing authority, which guaranteed the purity and fineness of the metal in the coin.

The Difference between a Coin's Face Value and Market Value. By imprinting its symbol on a coin, the issuing government publicly announced the piece's face value. This might be the same as the market value of the metal in the coin. More often, the coin's constant face value was different from the changing market value of the coin's metallic content. The issuing government deliberately set the face value above the market value, the difference being attributable to public confidence in the government. This difference had to be wide enough to prevent the market value from exceeding the face value. For if that happened, the coins would soon vanish from the market.

The Minter's Profit, Legitimate and Illegitimate. Because the issuing government had to set the face value of a coin above its market value, the government made a profit. This covered its expenses, for equipment, material, and wages, with something left over in normal times. But if the government suddenly faced financial difficulties due to war, poor harvest, civil unrest, and the like, it might seek a way out by covertly debasing the currency. The customary ratio of precious metal to base metal could be changed by lowering the silver content and raising the copper content. The size of the coin could be reduced. Both these tricks could be tried at the same time.

If the changes were so slight as to avoid detection by the population at large, the government would make an illegitimate profit. But the maneuver would fail, if the debasement of the currency were so gross as to be generally perceptible, with the debased coinage meeting resistance instead of acceptance.

Demonetization and Remonetization. As a coin passes from hand to hand, in the course of time it is worn down. When its silver content falls too low, it should be withdrawn as depreciated. If this condition becomes widespread, the whole issue should be replaced by a new coinage.

When the minting authority makes this decision, it should demonetize the old coinage. It should replace every old coin with an equivalent new coin. Permitting the two issues - the old and the new - to circulate side by side would create serious problems.

The minting authority might put in circulation a new issue pretending to have the same value as the old issue. But if the new issue were found to be inferior in quantity or quality, the reputation of the minting authority would suffer irreparable damage.

Nicholas Copernicus and Nicholas Oresme. Copernicus was not the first to write about monetary questions. Prominent among his predecessors was Nicholas Oresme (c. 1320-1382), who is generally regarded as the most brilliant mind of the fourteenth century. Oresme was in the service of the French king. That monarch had the power to enrich himself by debasing the currency at his subjects' expense. Such a misstep was the principal concern in Oresme's treatise on money.

Copernicus was a subject of the king of Poland. His most acute monetary problem was quite different from what confronted the king of France in Oresme's time. Oresme began his discussion of money with a quotation from the Bible. He also said: "In the present treatise I intend to write about what seems to me should be said, mainly according to the philosophy of Aristotle." In his monetary tracts Copernicus did not introduce a single quotation from Aristotle or from any of the ancient writers who were constantly cited by Oresme.

Did Copernicus Know about Oresme? Oresme's treatise on money is preserved in numerous manuscripts in French or in Latin. Two editions were printed before Copernicus first set down his thoughts about money in writing. Yet there is nothing to indicate that Copernicus was familiar with Oresme's treatment of monetary problems, whether in French or in Latin, whether in manuscript or printed form.

The same uncertainty exists in relation to astronomy. For, Oresme discussed the possibility that the earth moves in a way that has made some people think he was in some sense a pre-Copernican. But nobody has ever shown any dependence of Copernicus' astronomy on Oresme's. The same may be said about Copernicus' monetary writings. Whatever slight similarity may be found between the monetary treatises of these two writers may be due to their treatment of the same topic, which had been discussed since antiquity. Apart from their debt to these common sources, they diverge very widely, in astronomy as well as in monetary theory. It may well be asked whether Copernicus ever heard of Oresme.

Copernicus' Historical Method. For historical examples of monetary problems, Oresme looked back at the Bible and ancient secular writings. These concerned themselves with economic conditions quite different from what prevailed in Prussia and Poland in Copernicus' time.

The historical background of Copernicus' monetary discussions was provided by the Teutonic Knights. This military order conquered a vast area, which it ruled with an iron fist. It suffered a crushing defeat in Prussia on 15 July 1410 at the hands of Poland and Lithuania. The order's Grand Master, Ulrich of Jungingen, was slain on the battlefield, which lay between Tannenberg and Grunwald. The monetary consequences of this military disaster could be traced in the order's coinage.

As a member of the Varmia Chapter, Copernicus had access to the order's proclamations and correspondence. He was also able to handle the order's old coins. He weighed them and examined them. On the basis of this combination of documentary and numismatic evidence, Copernicus compiled the first history of Prussian coinage. He equipped himself thereby to deal with the pressing monetary problems of his own time.

The Debasement of the Prussian Coinage. Copernicus was able to compare Prussian coins issued before the battle of Tannenberg in 1410 with others issued thereafter. The later coins were made to look like those minted before the disastrous battle. But the silver content was reduced from 3/4 to 3/5. Once the Prussian coinage started down this slippery slope, the debasement grew steadily worse. What began as 3/4 silver + 1/4 copper became 1/4 silver + 3/4 copper. Such money should properly be called "copper coinage" rather than "silver coinage." Throughout this debasement, 112 shillings were minted from 1/2 pound of the alloy.

Good Money versus Light Money. A government made a serious mistake if it permitted two different currencies to circulate at the same time. When the coinage in people's hands was debased, the introduction of a good coinage would accomplish little, unless the inferior coinage was compulsorily withdrawn. The opposite occurred when the Teutonic Knights allowed the sound coinage, minted before 1410, to continue to circulate after they began to debase their money.

In an effort to rectify this situation, a new shilling was minted. On 7 July 1416 the Grand Master proclaimed that "two of the old shillings should be changed for one new shilling." The old shillings, however, remained in use. Hence, there were two kinds of marks. The new or good mark consisted of sixty new, good shillings. The old or light mark comprised sixty old, light shillings. Being much more numerous, the penny was not changed. The light shilling was still worth sixpence. But in accordance with the relation between the light shilling and the good shilling, the latter was equivalent to twelve pence.

Municipal Coinage. The Order of Teutonic Knights reserved to itself the exclusive privilege of coinage throughout its domains. When the Prussian towns and rural nobility revolted on 4 February 1454, they allied themselves with the kingdom of Poland, which they entered as Royal Prussia. The coinage of the Order, now an enemy state, had to be replaced. Polish money was not extended to Royal Prussia. Instead, on 9 March 1454 King Casimir IV decreed: "It is our wish that, only as long as the present war continues...money shall be minted in four places of the aforesaid lands [Prussia], namely, Toruń, Elbląg, Gdańsk, and Koenigsberg." Koenigsberg sided with the Teutonic Order. In 1457, on 15 May Gdańsk, and then on 26

August Toruń, were granted the minting privilege in perpetuity. The dismal history of the Order's coinage after Tannenberg continued under municipal auspices.

The Shilling, the Skoter, and the Groat. Because the new shilling had an increased silver content, it was rated equal to the skoter, traditionally worth 2 1/2 shillings. However, this skoter proved attractive to the western neighbors of Royal Prussia in the mark of Brandenburg. In an effort to stop this drain and draw the new shilling back, it was revalued by the Estates of Royal Prussia at a groat, traditionally 3 shillings. The Prussian groat was first minted by order of Johann of Tiefen, Grand Master from 1489 to 1497.

The Effect of Currency Debasement on Prussia's Foreign Trade. Prussia's active foreign trade, carried by boat overseas, by barge on the rivers, and by wagon overland, was threatened by the decline in the value of Prussian money. Copernicus foresaw a time when Prussia would lose its gold and silver. Its currency would be copper. Would a foreign merchant then be willing to export his goods to Prussia in exchange for copper coins? Would a Prussian merchant be able to pay for anything in foreign lands with Prussia's debased coinage? All the while its edges were being clipped. It was not until after Copernicus' death that the hammer-and-anvil process of minting began to be replaced by the mill-and-screw process as a safeguard against clipped edges.

The General Public Loses, the Expert Wins. When a currency in circulation consisted of coins debased in varying degrees, the ordinary person paid for his purchases without minutely examining the differences between the coins in his possession. But the expert was selective. He paid with inferior coins, and kept those of better quality. These he melted down, and sold the silver for more than the coins would have bought. Such destruction of the more valuable coinage was prohibited by law. But it was more difficult to enforce this law than to denounce the forbidden practice.

Copernicus Summarizes the Position of his Opponents. Copernicus was aware that his plea for a return to sound money was opposed by certain groups. They claimed that cheap money would make things easier for poor people. The cost of living would go down. Production of material goods would increase. These arguments were approved by the minters. For they made a profit from the existing situation. Any change might reduce their income or even put them out of business. In his own mind, an artisan priced his artifact on an absolute scale. As money became cheaper, he could easily raise his price. So could a merchant. Both artisan and merchant could therefore adjust to cheap money.

Copernicus' First Three Writings about Money. Copernicus was chosen Administrator of its holdings by the Varmia Chapter, to take office on 11 November 1516. His Administrator's duties brought him into close contact with the peasants. Among their many problems was the debasement of the coinage. In the earliest entirely empirical and secular discussion of the monetary problem, absolutely devoid of theological argumentation and quotations from the Bible, Copernicus wrote down his private *Meditations*, which he completed on 15 August 1517. Word of this Latin essay reached the Estates of Royal Prussia, which asked Copernicus to prepare a German version for them. He did so in 1519.

On New Year's Day, 1520, however, war broke out between the Teutonic Knights and Poland. On 5 April 1521 a ceasefire put an end to the hostilities. Poland seized this opportunity to press for a reform of the Royal Prussian currency that would make its various denominations readily equated with the denominations of the Polish currency. Such a proposal was added by Copernicus to the German version of his treatise that was read aloud to the Estates of Royal Prussia on 21 March 1522.

Copernicus' Fourth and Final Essay about Money. Three years later, a startling political development changed the framework of the Prussian monetary problem. The Order of the Teutonic Knights was dissolved. Its territory was converted into a secular state. Its former Grand Master became duke of (East) Prussia. This was ceded to the Polish king. He bestowed it on the duke as a fief owing allegiance to the crown. Article 28 of an agreement between the king and the duke, dated 8 April 1525, provided for the regulation of the currencies of the two Prussias, Royal and Ducal. This regulation was to be promulgated at a meeting to be held in the spring of 1526. If Copernicus was to be able to influence the outcome of that meeting, he had to take cognizance of recent developments, and recast his thinking and proposals. The outcome of that revision was his famous *Essay on the Coining of Money.*

The Desirability of a Single Mint. The incorporation of the former territory of the Teutonic Knights as the duchy of (East) Prussia within the kingdom of Poland made it possible to eliminate a troublesome feature of the coinage circulating in Ducal Prussia and Royal Prussia. The four mints operating in both Prussias produced coins of the same denominations. Although these coins struck by the four mints bore the same designations, some people wondered whether they were cast from identical alloys. Did they have identical sizes and weights? Doubts of this kind were a hindrance to the smooth functioning of business transactions.

In his *Essay* Copernicus stressed the importance of having only one mint serve both Prussias. The coins produced by that single mint would bear the insignia of Royal Prussia on one side, and of Ducal Prussia on the other side. On both sides the Prussian insignia would be surmounted by the crown to symbolize the overlordship of the Polish kingdom.

Two Prussian Mints as a Concession to Practical Politics. The duke of Prussia could be expected to demand the right to retain his own mint in Koenigsberg. But of the three mints in Royal Prussia, two should be suppressed. These two municipalities would undoubtedly protest vigorously against their loss of income. However, the single mint in Royal Prussia should be managed, not by the local municipality, but by a royal appointee. He should not seek to make the mint profitable. He should lower the copper content in the alloy to the point where it would cover the expenses of the minting process, and no more. His coins should be uniform with those produced by the ducal mint. Both Prussian coinages should pass freely throughout the entire Polish kingdom. Smooth business transactions between Prussian tradesmen and their counterparts in the rest of the kingdom would reduce mutual mistrust, and encourage economic activity.

No Remonetization without Demonetization. When Copernicus wrote his *Essay,* the Prussian monetary system was in a confused state. Within the various denominations, coins of unequal intrinsic value were in circulation. To add a new coinage to this hodgepodge would only make matters worse.

Once the new monetary system has been devised, and the new coins struck, stated Copernicus, issue them at the mint in exchange for the old coins, rated at their intrinsic worth. Withdraw all the old coins from circulation completely and quickly.

Pay no attention to those who urge that the old coins should be retained and rated in terms of the new coinage. This would be a hopeless task, since so many different kinds of coins were circulating in the several denominations down to the penny.

Copernicus was aware that anyone who surrendered his old coins at the mint would suffer a loss when he received the equivalent of their intrinsic worth in the new coinage. But he would lose only that one time. What he gained would be far greater and last longer: a stable monetary system permitting increased production and a rising level of prosperity.

How Copernicus Contributed to the Establishment of the New Prussian Currency. The details of the renewed Prussian monetary system were to be decided at a meeting of the Estates of Royal Prussia. As a canon of Varmia, Copernicus was not himself a member of the Estates. But the bishop of Varmia was by prescriptive right the Estates' presiding officer. When the coinage question was on their agenda, the bishop could ask the Varmia Chapter to let Copernicus accompany him as one of his advisers. In 1522 the Estates had invited him to explain to them in their own language his views about money. But on this occasion in 1526 it is uncertain whether Copernicus was present when the Estates met in Gdańsk to hear about their new money. They were, however, thoroughly familiar with his *Essay,* in which he set forth a practical program regarding the new coinage.

The New Prussian Coinage in the Rest of Poland. A proposed solution of the problem of the new Prussian coinage fulfilled only part of what Copernicus undertook in his *Essay.* The use of that new money was not going to be confined to Royal Prussia and Ducal Prussia. It was authorized to circulate freely throughout the rest of the kingdom of Poland, which was accustomed to Polish money. Was there some convenient way for ordinary people to accept an integral number of Prussian coins as equivalent to an integral number of Polish coins?

In order to evaluate these two coinages, Copernicus turned to a common measure, the precious metals. Silver provided the computational basis for the money of Prussia, which lacked the raw material to mint gold. The gold coins circulating in Prussia came in from elsewhere. Of these foreign gold coins, the finest was the Hungarian florin. This was so nearly pure that it was exchangeable for its weight in gold bullion. The Hungarian florin's exchange rate against Prussian currency and Polish currency permitted Copernicus to match these two currencies with each other.

The Interchangeability of Prussian and Polish Money. A Prussian mark was divided into 20 Prussian groats. Outside Prussia, the Prussian mark was worth 20 Polish groats, so that 1 Prussian groat = 1 Polish groat. This convenient interchangeability lasted as long as the ratio of silver to gold was 11:1, that is, while 1 lb. silver was worth 1/11 lb. gold. From that quantity of gold, 10 Hungarian florins were minted. From 1 lb. silver, 20 Prussian marks were made. Hence, 1 Hungarian florin equaled 2 Prussian marks or 40 Polish groats.

But before Copernicus wrote his *Essay,* the value of silver fell somewhat, to the point where its ratio to gold became 12:1. Hence, 20 Prussian marks were no longer worth 10 Hungarian florins. Nor was 1 Prussian mark (= 20 Prussian groats) equal to 20 Polish groats. If the Prussian mark were devalued (say, from 20 to 15 Polish groats), the simple equivalence of a Prussian groat with a Polish groat would be ended.

Copernicus' Sixfold Program. Copernicus' *Essay* was more than a historical survey of Prussian coinage, more than a theoretical investigation of the monetary problem, more than a lament for the uninformed public, and more than an indictment of the avaricious goldsmiths. It was also a program for planning the new Prussian currency, valid in Ducal Prussia as well as in Royal Prussia, and conveniently interchangeable with Polish money.

Copernicus summed up his monetary ideas in the following six recommendations. The unanimous consent of the Estates of Royal Prussia should be obtained for any proposed action. If possible, only one mint should be authorized to serve both Prussias. The currency in circulation must be withdrawn as the new currency is issued. The interchangeability of the Prussian groat with the Polish groat should be retained, if feasible. Avoid an excessive number of denominations. Issue all the new denominations at the same time. Each one of such decisions should be taken, not merely with the current crisis in mind, but with a view to the long-range stability of the economy.

The Royal Proclamation. On 8 April 1525 the king of Poland promised the duke of Prussia that he would convene an official meeting by 20 May 1526 to settle the question of the new Polish currency. The session took place in Gdańsk, where the royal decree was issued on 17 July 1526. Its specific provisions show that Copernicus' *Essay* harmonized with the advice given to the king by his financial advisers: simultaneous remonetization and demonetization; royal and local insignia on the coins; only three denominations; and acceptance of Prussian money throughout Poland.

Copernicus' emphasis on the interchangeability of the Prussian groat with the Polish groat was extended by the decree to the penny: 1 Prussian penny = 1 Polish penny.

In the area of foreign exchange and the ratio of silver to gold, the decree upheld the equivalence of 2 Prussian marks to 1 Hungarian gold florin.

The Recognition of Copernicus as an Authority on Monetary Questions. In 1526 the king of Poland issued two decrees concerning the new coinages. His regulations governing Prussian money were proclaimed in Gdańsk on 17 July, and those dealing with Polish currency were promulgated in Cracow on 15 October. But administrative problems connected with the Prussian money still remained to be solved by the Estates of Royal Prussia. For such a meeting, scheduled to be held on 16 March 1528, the Varmia Chapter selected two of its members to be its spokesmen. A week before the session was to begin, however, the bishop of Varmia notified the Chapter that he was designating Copernicus to be his adviser. In the same letter he asked the Chapter to add Copernicus to the two canons already chosen to represent it.

At the meeting of the estates in mid-March 1528, the delegates of Ducal Prussia admitted that they still lacked complete authority. Soon thereafter it became known that they were being clothed with plenipotentiary power. The bishop of Varmia hurriedly requested that the Varmia Chapter should arrange to have Copernicus leave at once for Lidzbark. There he and the bishop could go over the currency question in preparation for the Estates' next meeting in May 1528.

The bishop was deeply impressed by Copernicus' performance in these preparatory talks in Lidzbark. When the astronomer returned to Frombork, he delivered a letter from the bishop. This requested that the canons, acting officially as the Varmia Chapter, should choose Copernicus to be their spokesman and the bishop's adviser at the Estates' forthcoming meeting in May 1528.

Evaluating the Old Coins in Terms of the New Coins. At that time an evaluation commission was created, comprising representatives from both Ducal Prussia and Royal Prussia. The latter's seven delegates included Copernicus. The process of replacing the old coins was controlled by the speed with which the new coins were produced. As a result, various kinds of money were still being passed from hand to hand. The purpose of the commission was to evaluate, in terms of the new money, every denomination of silver and gold coinage still in circulation.

The Royal Mint in Toruń. The Polish government decided that the king's mint in Royal Prussia would be located in Toruń. It could begin operations without delay, the king's secretary informed the Estates of Royal Prussia at their meeting in May 1528, if they would reach an agreement about the issuance of the new Prussian coinage. But the Estates postponed action until their next meeting, in July 1528.

This session was held in Toruń. The royal secretary, who had written from Cracow to the Estates on 28 April 1528, had meanwhile been appointed on 15 June 1528 master of the royal mint in Toruń.

The First Master of the Royal Mint in Toruń. Justus Ludovicus Decius was born in Wissembourg (Weissenberg). His father, the mayor of this Alsatian town, was unable to control the devastating private feuds. Hence, young Dietz (as he was called before he latinized his surname) left, as did many others of his age. He tried his luck here and there, and finally he went to work for the king's banker in Cracow. Through this employment he gained the royal favor, and rose to be the king's secretary. His banking experience qualified him to take a prominent part in the projected reform of the currency. When the decision was made to establish the royal mint in Toruń, Decius was appointed master.

Copernicus Asks Decius for New Pennies for the Varmia Chapter. A decision to withdraw the old pennies from circulation was adopted by the Estates of Royal Prussia on 23 July 1528. On that basis Decius, as master of the Toruń mint, struck 6000 marks' worth of new pennies. These could be obtained by the members of the Estates, with payment to be made later on in old pennies. Copernicus did not attend the meeting of the Estates in July 1528. But he was present at the meeting in February 1529. At that time, on behalf of the Varmia Chapter, he asked Decius for 40 marks' worth of the new pennnies, which the bishop paid for with old coins.

Copernicus Is Asked to Be the Voice of the Varmia Chapter. The bishop of Varmia regarded the coinage problem as the most complicated he faced. As the monetary discussions in the Estates of Royal Prussia dragged on and on, he relied more and more on Copernicus' knowledge and insight. Thus, on 27 April 1529 the bishop wrote to the Varmia Chapter about the Estates' next meeting. This was scheduled to begin on 8 May. He reminded the canons that they were familiar with the whole problem from its very beginnings. He wanted them to consult among themselves, reach conclusions about what should be done at the forthcoming meeting of the Estates, and give complete instructions to Copernicus as the intermediary. He need not arrive on the day the session would begin. Several days later would be better. But he should convey the Chapter's decisions to the bishop, and advise him in this and all other matters to be taken up by the Estates.

Copernicus Is Chosen to Represent the Varmia Chapter on the Currency Commission. The discussions of the problems connected with the new coinages dragged on and on. Each special interest sought its own advantage, in disregard of the general welfare. The Estates of Royal Prussia proved to be too large and too unwieldy a deliberative body to settle the numerous differences of opinion fairly and amicably. Hence all sides agreed to create a smaller currency commission.

While on his way back to his residence in Lidzbark, the bishop of Varmia conferred with the canons in Frombork. He appointed two commissioners, one of whom was Copernicus. The other, Felix Reich, happened not to be in Frombork at the time. Hence the bishop notified him of his appointment by a letter written on 15 October 1530.

Reich Declines to Accept his Appointment to the Currency Commission. When the bishop's letter reached Reich, he was engaged in compiling the yearly report of his financial activities. This had to be ready for presentation to the Chapter at its general meeting early in November. Reich explained in a (lost) letter to the bishop that it was impossible for him to prepare for the session in Elbląg, go there, participate in the discussions, and return in time for the Chapter's annual review. In that case, the bishop wrote to the Chapter, it should proceed to elect someone else to represent it, along with Copernicus.

The Necessity of a Second Delegate in Addition to Copernicus. Meanwhile, several canons called on the bishop to tell him that Reich would not be able to serve on the currency commission. In a letter they proposed a number of different solutions. But the bishop could not pick a replacement for Reich. Because he was not on the scene, he did not know who was too sick to go, or otherwise impeded. He insisted, however, that Copernicus should not be the only representative of the Chapter on the commission. For if the Chapter had a second delegate, it might obtain the commission's chairmanship, a post of considerable authority. The Chapter must be on guard that nothing detrimental to itself or to him should be done by the commission.

Copernicus Differs with Decius on How to Evaluate Gold Coins. In place of Reich, the Varmia Chapter elected Alexander Scultetus to accompany Copernicus to Elbląg. On 29 October 1530 Decius delivered a long speech against excessive evaluation of gold coins, on the ground that it would harm the poor. On the following day Copernicus argued that the market price of gold and silver bullion by weight should be ascertained before individual coins of undetermined alloy could be evaluated. Copernicus' insistence on the necessity of prior market research seemed impractical to the commission's other members. They wanted immediate results, but disagreed about what they should be. Copernicus stayed away from the subsequent debates.

Decius' Warning that Currency Negotiations Must Be Kept Secret. When the Estates of Royal Prussia were summoned to discuss the monetary problem, Decius sent them his proposals to be approved or disapproved. He emphasized the necessity of absolute secrecy. Those who learned about what was projected could take advantage of their prior knowledge. They could withdraw the good coins, to the detriment of the community at large.

Copernicus and the *Letter to Decius.* The king's monetary edict was issued on 17 July 1526. On the following day a reply was sent to Decius in the name of the Councillors of Prussia from their public meeting in Gdańsk in 1526. No individual was specified as the author of this *Letter to Decius.*

Later on, however, a confusion arose when handwritten copies of Copernicus' *Essay* and the *Letter to Decius* were included in a composite manuscript. The *Letter to Decius* was misascribed to Copernicus. It is not even certain that he was physically present in Gdańsk during the meeting of the Estates there in 1526. His economic ideas, however, were certainly present, in the minds of those who wrote the *Letter to Decius.*

That document repeats a number of Copernicus' conceptions. Sometimes its language echoes his. In one respect its authors emphasize something of importance that was not said explicitly by Copernicus: the price of silver bullion had risen steeply when paid for in depreciating currency, but fell much less when paid for in gold.

Copernicus, not Gresham. "Gresham's Law," explaining the disappearance of good money when it is circulating alongside bad money in the market place, was misattributed over a century ago to Sir Thomas Gresham (c. 1519-1579). In his role of monetary adviser to Queen Elizabeth I of England, Gresham deplored the loss of good coinage overseas. He was also aware that all the bad money remaining in England was not equally bad. Some coins were better than others. Yet, the better coins were not driven out of circulation. For they were scarce enough to be needed for business purposes. *(See Reading No. 37.)*

In the absence of "Gresham's Law," Copernicus' writings on money contain the earliest expression of the principle in a formal treatise. Of course, he was not the first to observe the phenomenon. Many others were painfully aware of it, but they did not compose essays analyzing it and proposing remedies. If they did, their output has not yet come to light.

CHAPTER 8

COPERNICUS, ASTRONOMER

Cardinal Bessarion. After a rift of many centuries, under the increasing pressure of the Muslim Turks, the two major Christian churches - the Greek Orthodox in the east and the Roman Catholic in the west - were contemplating a reunion. For this purpose the Council of Florence was convened in 1438. Among the Greeks, the foremost proponent of reunion was Archbishop Bessarion of Nicaea. As an individual, he abandoned Orthodoxy and embraced Roman Catholicism, which made him a cardinal.

Cardinal Bessarion incessantly urged a crusade against the Turks. In intellectual matters, he was deeply distressed that educated people in the West knew so little about the literature of ancient Greece. Some of his compatriots stressed the teaching of the Greek language to their new associates. Bessarion preferred having the Greek classics translated into Latin. Although he himself did not begin to learn that language until he was a mature man, he quickly won recognition as "the best Latinist among the Greeks."

Bessarion and Ptolemy. Bessarion knew that in the field of astronomy Ptolemy stood head and shoulders above all the Latin writers, and the Greek as well. Hence, the cardinal asked one of his agents to procure a Greek manuscript of Ptolemy's *Syntaxis* for him as quickly as possible. As soon as it arrived, he started to translate it into Latin. However, the time he could devote to this task was severely limited by the burden of his official duties.

Bessarion in Vienna. As a papal legate, Bessarion arrived in Vienna on 5 May 1460. It was his mission to encourage the Holy Roman Empire to join in taking military action against the Turks. On the educational front, the cardinal found that the University of Vienna offered no instruction in Greek. Professor George Peurbach, however, realized what his students were missing without access to Ptolemy. For centuries, university teaching of astronomy had been conducted on a much lower level than Ptolemy's. That level could be raised by a satisfactory Latin version of Ptolemy. An unsatisfactory translation into Latin had been made about three hundred years earlier. This was based on an Arabic translation from the Greek, not on the original Greek text itself. In his own minute handwriting, Peurbach copied the medieval Latin translation. He made many annotations, the product of his constant reading and re-reading of the translation. He had gone over it so many times that he knew it almost by heart. Yet, he realized it was so bad that "not even Ptolemy himself, if he were restored to life, would accept it as his own work."

Peurbach's Ptolemy. Bessarion advised Peurbach, instead of translating Ptolemy, to "make him briefer and clearer." The cardinal invited the professor to accompany him on his return trip from Vienna to Italy. Peurbach accepted, provided that his favorite pupil, Johannes Regiomontanus (1436-1476), was included in the invitation. Bessarion agreed. Peurbach pressed on with his *Epitome* of Ptolemy. In his late thirties on 8 April 1461, he died after doing only the first six of the thirteen Books into which Ptolemy had divided the *Syntaxis.* As Peurbach was passing away, to Regiomontanus, who was holding him in his arms, he spoke memorable words. *(See Reading No. 38.)*

Regiomontanus' Completion of Peurbach's *Epitome.* After the death of his beloved teacher, what Peurbach had begun in Vienna was finished in Italy by Regiomontanus. In the *Epitome,* their joint product, the disciple's contribution differs from the master's in two important respects. Peurbach had never learned Greek. On the other hand, by close association with Bessarion and his Greek-speaking circle in Italy, Regiomontanus acquired the language of Ptolemy. Instead of being confined, like Peurbach, to the Latin *Syntaxis* that had been translated from Arabic, Regiomontanus could consult Bessarion's Greek manuscripts of Ptolemy's astronomical masterpiece.

Ptolemy's *Syntaxis* in a Latin Condensation. Regiomontanus had a second great advantage over Peurbach. Ptolemy's Greek *Syntaxis* in thirteen Books had been condensed into six Books in Latin. Who did this condensation is not known. He gave it no definite title. Like other manuscripts without author and title, it may be identified by its incipit or opening words: *Omnium recte philosophantium.* Regiomontanus made a copy of it for his own use, and added many annotations.[78] His copy is still preserved.

Regiomontanus' Copy of the Condensation: Full Title and Short Title. *Omnium recte philosophantium* was an influential manuscript, of which several copies survive. Latin writers who cited it had to devise a title by which they could refer to it. In Regiomontanus' copy, it is entitled *Almagesti minoris libri VI* = Six Books of the Abbreviated Almagest (the hybrid Arabic misnomer of the *Syntaxis).* Once this full title has been introduced to a modern reader, a shorter title may be desired. If so, the last two words *(libri VI)* are dropped. Then the remaining first two words *(Almagesti minoris)* are in the genitive or possessive case. Such a floating genitive, with nothing to depend on, is unacceptable as a short title. In English, we would not tolerate "Kennedy's" as a short title, while suppressing, say "Letters." Similarly, *Almagesti minoris* is unsatisfactory as an independent title. This expression in the genitive case must therefore be shifted to the corresponding nominative case, beginning with *Almagestum.* Since this belongs to the neuter gender, it should be accompanied by the neuter adjective *minus.* Yet, a recent article on Peurbach couples "Almagestum minor," thereby violating an elementary rule of Latin grammar by combining a neuter noun with a masculine/feminine adjective.[79] This is a joint article, finished by a younger contibutor after the death of an older colleague, just as the *Epitome* was completed by Regiomontanus after Peurbach's demise. The ungrammatical combination "Almagestum minor" was introduced previously by the younger contributor elsewhere.[80]

Regiomontanus' Dedication of the *Epitome.* When Regiomontanus finished the *Epitome,* a professional scribe made a calligraphic copy on parchment of the entire work, which was dedicated to Bessarion. That parchment manuscript still survives, in the library of St. Mark in Venice. In this parchment, when Regiomontanus addressed Bessarion, he gave him his various titles, but he did not call him "Patriarch of Constantinople." That former capital of the Byzantine Empire and of the Orthodox Christian Church had been captured by the Turks in 1453. Thereafter the Roman Catholic Church appointed a nominal or titular Patriarch of Constantinople. Bessarion received that honor shortly after the death of his predecessor on 27 April 1463. Before that month was out Bessarion was named Patriarch of Constantinople. But he had not yet been elevated to the patriarchate when Regiomontanus presented to him that magnificent parchment manuscript of the *Epitome.* Hence Regiomontanus completed his post-Peurbachian part of the *Epitome,* and dedicated the whole volume to Bessarion, before the end of April 1463.

Bessarion Did Not Print the *Epitome.* Bessarion deserved to be honored as the dedicatee of the *Epitome.* He had encouraged Peurbach to work on it. He had taken Regiomontanus with him to Italy. There they both agreed to reject a fresh Latin translation of the *Syntaxis* from the Greek by a Greek because they found it defective technically. Bessarion was highly enthusiastic about promoting knowledge of the Greek classics through the medium of excellent Latin treatises like the *Epitome.* He was immensely wealthy. Between his receipt of the *Epitome* and his death on 18 November 1472, about a decade elapsed. In all that time he lifted not a finger nor spent a penny for the printing of the *Epitome.* Why?

Regiomontanus Becomes a Publisher in Nuremberg. Bessarion and Regiomontanus parted company about 1465. After serving as scientific adviser to the king of Hungary, in 1471 Regiomontanus informed a correspondent that he had chosen Nuremberg as his permanent home. *(See Reading No. 39.)* In Nuremberg, Regiomontanus became the first publisher of astronomical and mathematical literature. Such writings were avoided by previous printers. They were reluctant to invest their capital in works involving diagrams, which required special craftsmen and additional expense.

Regiomontanus Planned to Publish the *Epitome.* If a publisher is going to stay in business, he must be able to sell what he prints. Especially if he is a specialist, like Regiomontanus, he cannot rely exclusively on the potential customers in his immediate vicinity, even if that vicinity is as prosperous a manufacturing and commercial center as Nuremberg was. Regiomontanus was astute enough as a businessman to realize that he had to resort to advertising. With this purpose in mind, he designed and distributed as widely as possible what is now known in the trade as a "broadside," a single sheet of paper printed only on one side.

Regiomontanus' broadside listed some works already printed, others "almost finished at the present time," and forty-one others still to be printed. The last group included the summary of Ptolemy. Here it was entitled, not *"Epitome,"* but "Breviarium," a Latin equivalent of the Greek word *"Epitome."* Hence, when Regiomontanus gave Bessarion a professional scribe's copy of the *Epitome* on parchment, he retained his own copy. This is what he announced in his broadside that he intended to print. His untimely death in 1476 prevented him from doing so.

The Printing of the *Epitome* after a Third of a Century. Regiomontanus died at the early age of forty while he was in Rome at the behest of the pope to try to improve the ecclesiastical calendar. His manuscript of the *Epitome* would not have been of much help for this purpose. He did not transport his entire personal library to Rome, where he expected his stay to be temporary. In all probability, then, his manuscript of the *Epitome* remained behind in Nuremberg. It has not survived.

Bessarion's copy, on the other hand, was preserved in the library of St. Mark in Venice, which received the nucleus of its collection from the wealthy cardinal. His presentation copy of the *Epitome* lay undisturbed until an alert and energetic editor of astronomical texts began to make waves in the quiet Venetian lagoons surrounding its resting place. He published the second printed edition of the *Alfonsine Tables.* In the dedication, dated 31 October 1492, he expressed his grief that Regiomontanus' wonderful achievements had been neglected. He announced that he was then "striving to bring out, in a thoroughly corrected form, Johannes Regiomontanus' Breviarium [that is, *Epitome*] of Ptolemy under his own editorship and auspices."

For reasons which are not yet known, he failed to do so. On 10 February 1496 another publisher submitted a petition for the customary ten-year copyright. The Venetian authorities granted it. On 31 August of the same year the *Epitome,* begun more than a third of a century earlier by Peurbach, and finished fairly promptly by Regiomontanus to keep his promise to his dying teacher, finally emerged from the obscurity in which it had lain hidden so long. *(See Reading No. 40.)*

The *Epitome* and Copernicus. Not long after the publication of the *Epitome* in Venice, Copernicus arrived in Bologna. As a student in a foreign country, did he have money enough to buy such an expensive book? No copy of the *Epitome* bearing the signature of Copernicus as owner has ever been traced. Yet it exerted a profound influence on his development as an astronomer. In all likelihood he managed to consult a copy owned by a friendly fellow astronomer.

Copernicus' *Commentariolus*. The first astronomical essay written by Copernicus left his hands without a title on it. After his death it was christened *Commentariolus.* This "little commentary" is of unique importance in the history of human thought. It was the earliest presentation in a carefully reasoned form of the true cosmic status of the earth as a planet of the sun. Without saying so explicitly, it contradicted mankind's immemorial belief that the earth is stationary. This popular notion had long been supported by elaborate arguments of philosophers and astronomers, foremost among them Ptolemy.

A Defect in Ptolemy's Lunar Theory. As the moon revolves around the earth, its distance from the earth varies. As a result, its apparent diameter varies in length from 33'30" to 29'21', or from 1 1/7 to 1. In Ptolemy's lunar theory, the distance between the earth and the moon was reckoned as a multiple of the length of the earth's radius. About 33 1/2 earth radii separated the earth from Ptolemy's moon when it was at its perigee, or closest approach to the earth (*Syntaxis,* V, 13, 15). At the apogee, or greatest distance from the earth, Ptolemy's moon was over 64 earth radii away. Then the diameter of Ptolemy's moon should look nearly twice as long at perigee as at apogee. Ptolemy's ratio of almost 2:1 is very different from nature's 1 1/7:1.

This Defect in Ptolemy's Lunar Theory Was Pointed Out in the *Epitome*. The *Epitome* (Book V, Proposition 22) put the apogeal distance of Ptolemy's moon at 64 1/6 earth radii, and its perigeal distance at 33 1/2 earth radii. At half-moon, the lunar disk is half visible, half invisible. Assume the half-moon to be perigeal. If the disk, half of which is seen, were imagined to be visible in its entirety, it would look four times the size of the full moon. This is a necessary consequence of Ptolemy's lunar distances. But it never occurs in nature. This defect in Ptolemy's lunar theory was pointed out in the *Epitome. (See Reading No. 41.)* In the *Commentariolus,* Copernicus repeated the *Epitome's* criticism of Ptolemy's lunar theory in his own words.

The *Commentariolus*' Second Objection to Ptolemy's Lunar Theory. Without mentioning Ptolemy by name, the *Commentariolus* showed that his lunar thoery made the moon look bigger than it actually does. This undermining of the infallibility of the "prince of astronomers," as he is called in the *Epitome's* very first line, was undoubtedly derived by Copernicus from the *Epitome,* V, 22. The

Commentariolus' second objection to the Ptolemaic lunar theory, however, is not found in the *Epitome*. The *Commentariolus* charged that the lunar theory under scrutiny "improperly treated the motion on the eccentric as nonuniform." This rejection of nonuniform motion recalls the condemnation of the equant, near the beginning of the *Commentariolus*, as a device that failed to satisfy astronomy's theoretical requirements, as then understood. *(See Reading No. 42.)*

Ibn al-Haytham's Denunciation of Ptolemy. Al-Hasan Ibn al-Haytham (965-c. 1040), one of the greatest medieval Islamic scientists, was known in the Latin West as Alhazen. Late in his career he developed a critical attitude toward Ptolemy. He composed a work entitled *Doubts about Ptolemy*, usually cited as *Shukuk*, an Arabic word meaning "difficulties." In his *Shukuk*, Ibn al-Haytham "asserts the fallibility of men, even great men, and prescribes criticism of one's own views and those of others as the only method of advancing scientific knowledge."[81] Ibn al-Haytham coupled this advocacy of criticism in general with a repudiation of Ptolemy in particular. The Muslim scientist charged that Ptolemy was aware of the unsoundness of his views. These must be replaced by a true theory according to Ibn al-Haytham, who failed to supply it. *(See Reading No. 43.)*

Ibn al-Haytham and Proclus. The Islamic traditionalist's position that Ptolemy's authority should not be questioned was opposed by Ibn al-Haytham. He held that "it was appropriate for a Traditionalist to follow authority, but nothing less than a demonstration should satisfy the mathematician."[82] In some respects Ibn al-Haytham's critical attitude toward Ptolemy recalls the *Hypotyposis* of Proclus. His writings were popular among some intellectuals of the Byzantine Empire. A considerable quantity of their literature in Greek was translated into Arabic, and diffused to Islamic centers of learning. In the course of this transmission, did Proclus' *Hypotyposis*, or some echo of it, reach Ibn al-Haytham?

His earlier purpose was to expound Ptolemy. Later on, he became critical of the Greek scientist. Was this shift a result of Ibn al-Haytham's own intellectual development? Or was he also affected by Proclus or some follower of that widely read Greek thinker?

The ultimate outcome of Proclus' analysis of the prevailing astronomical theories was a conciliatory compromise. The astronomer's mathematical devices, while imaginary, were useful for the purposes of computation. Some were better than others. But even the best need not exist. Ibn al-Haytham was more forthright. Ptolemy's arrangement was merely imaginary. He knew that it was false. The search must go on for the truth.

Did Copernicus Know about Ibn al-Haytham's *Shukuk?* The search for the truth was continued by Copernicus. Did he know about Ibn al-Haytham's denunciation of Ptolemy in the *Shukuk?* No evidence indicates that he did, or that he ever heard of Ibn al-Haytham.

On the other hand, Copernicus knew something about what had been done in astronomy between Ptolemy's time and his own. He derived this information from the *Epitome.* It contained much material about post-Ptolemaic astronomers, mainly Muslim. This was presumably drawn from the "Abbreviated Syntaxis" or *Almagestum parvum,* as it was often cited, or *Almagesti minoris libri VI,* in Regiomontanus' copy. Was Copernicus' knowledge of post-Ptolemaic astronomy limited to what he found in the *Epitome?* No other source has as yet been identified.

Copernicus' Debt to the *Epitome* and Bessarion's Objection to it. What Copernicus learned from the *Epitome* about Muslim and other astronomers since Ptolemy loomed large in his outlook on long-term phenomena. This post-Ptolemaic information (and misinformation) deeply affected his thinking. But what he found interesting and important in the *Epitome* was profoundly disturbing to Bessarion.

Cardinal Bessarion's main purpose in life was to fight against Islam. Military means came first, but intellectual weapons were not to be despised. The *Epitome* should have been, in his view, a Latin condensation of the *Syntaxis* that would bring Ptolemy's astronomy within the reach of a wider reading public. Instead, in V, 22, it called attention to a flaw in Ptolemy's lunar theory. What was worse, it went on and on about astronomers of Islam. That should be exterminated, not extolled. Hence, Bessarion did not pay to have his presentation parchment copy of the *Epitome* printed.

Regiomontanus, on the other hand, was primarily devoted, not to the military destruction of Islam, but to the advancement of astronomy. If a scientist accomplished anything noteworthy, it went into the *Epitome,* whether the notable was a Christian, Muslim, or Jew. No wonder Regiomontanus and Bessarion parted company not long after the completion of the *Epitome* and Bessarion's decision not to pay the cost of having it printed.

The Sun's Apogee, Fixed or Movable? Before Copernicus, astronomers believed that the sun revolved around the earth once a year; that in the course of the year the distance between the sun and the earth varied; that, consequently, the earth was not in the center of the circle on which the sun moved; and that the sun reached a point on the circumference it traversed where it was at its farthest, or apogee, from the earth. The sun's apogee was found by Hipparchus to be $65^{\circ}30'$ from the vernal equinox. Nearly three centuries later, the same determination was made by Ptolemy. He therefore concluded that the solar apogee was fixed. This is what Copernicus repeated in his *Commentariolus.* Later on, however, he found out that the solar apogee or apse moved. Presumably that motion would always proceed in the same direction, eastward.

The *Epitome*'s Misinformation about the Solar Apogee. According to the *Epitome*, an expert Islamic astronomer reported finding the solar apogee 4°27' west of a predecessor's determination. The later astronomer, al-Zarqali (or Arzachel, as he was called in Latin; †1100) was reported by the *Epitome* to have been "therefore compelled to say that the center of the sun's eccentric moved on a certain small circle." With the solar apogee fixed, as in Ptolemy, there was no need for the center of the sun's eccentric to move. But if the solar apogee moved, sometimes eastward and sometimes westward, that result could be achieved by mounting the center of the sun's eccentric on a small circle. *(See Reading No. 44.)*

How Copernicus Was Baffled by the *Epitome*. Some of Copernicus' critics have complained that he merely followed his predecessors' observations slavishly. Yet, in the matter of the solar apogee, he was confronted by an anomaly. According to the findings of two highly renowned Muslim astronomers, as presented by the *Epitome*, the solar apogee moved backward. This would have been a unique occurrence in the long history of the determinations of the solar apogee. Apart from this exception, "The apogee appeared in a continuous, regular, and progressive advance until the present time," as Copernicus put it. The exception was due to a mistake, he said. *(See Reading No. 45.)*

Al-Battani on the Solar Apogee. Like Peurbach and Regiomontanus, the authors of the *Epitome*, Copernicus did not know Arabic. What the *Epitome* called strange, he recognized was a mistake. He was in no position, however, to correct the mistake. For instance, he had no access to a Latin translation of al-Battani's monumental work, usually called *Zij* in Arabic. From a Latin translation, the *Epitome* (or its source, the *Abbreviated Syntaxis*) extracted the essence of what al-Battani had written about the solar apogee. *(See Reading No. 46.)*

Al-Zarqali on the Solar Apogee. The *Epitome* could not extract the essence of what al-Zarqali had written about the solar apogee because his treatise on the motion of the fixed stars was not translated into Latin. From the Arabic original, which has not been preserved, only a translation into Hebrew survives in a single manuscript. This has been printed, together with a translation into Spanish. In an English version of this modern Spanish translation of the medieval Hebrew translation of the lost Arabic original, al-Zarqali said: "It is known that the sun's apogee advances among the fixed stars and in the order of the zodiacal signs [eastward] 1° in the period of 279 common years."[83] Hence, al-Zarqali did not believe that the solar apogee ever moved westward or backward.

With regard to the solar apogee, then, what the *Epitome* reported about al-Battani is absolutely correct. On the other hand, what it said about al-Zarqali is all wrong.

Why the Epitome Went Astray with Regard to Al-Zarqali. Without access to al-Zarqali's treatise on the stars, the *Epitome* (or its source) relied on a Latin translation of the *Toledan Tables*. The Arabic original of this composite work is lost. Some of its tables were written by al-Zarqali. His name was gradually linked more and more prominently with the *Toledan Tables*, of which he came to be regarded as the author. He had lived in Toledo until repeated attacks by the Christians drove him farther south in Spain. As regards the solar apogee, the *Toledan Tables* located it at two signs, 17°50'.[84] A zodiacal sign being equal to 30°, this position amounted to 77°50'. Without further ado, the *Epitome* (or its source) attributed 77°50' to al-Zarqali, without indicating that the *Toledan Tables* measured celestial motions from a fixed star.[85]

With and without the Precession of the Equinoxes. The precession, or westward motion of the equinoxes, was discovered by Hipparchus. He put the precessional rate at no less than 1/100° in a year. In agreement with Hipparchus, in the *Syntaxis* (VII, 2, end) Ptolemy accepted the rate as about 1° in 100 years. Al-Battani's value was somewhat faster, 1° in 66 years. He located the solar apogee at 82°17' from the vernal equinox. But the *Toledan Tables'* position of 77°50' was measured, not from the movable equinox, but from a fixed star. If the appropriate precession is added to the Toledan position, it becomes east, not west, of al-Battani's. Then the unique exception in the history of the determinations of the solar apogee disappears. The *Epitome* was in error with regard to al-Zarqali, as Copernicus recognized. But he failed to understand why the *Epitome* erred. The correct explanation was finally provided by the recent investigator of the *Toledan Tables* who was cited in note 85, above. He also cleared up the old misunderstanding concerning the number of al-Zarqali's observations of the sun in connection with the equinoctial and solstitial points.

The Number of al-Zarqali's Solar Observations. In the *Syntaxis* (III,1) Ptolemy admitted that "in general the observations of the solstices are hard to determine." The *Epitome* (III, 14, beginning) explained that "the sun's entry into the solstitial points cannot be obtained without great difficulty, because the sun's declination [distance in degrees and minutes along a circle perpendicular to the celestial equator] varies minimally in that region." Hence, Islamic astronomers proposed to overcome the difficulty of pinpointing the solstice by observing the sun's entry into the midpoints of four zodiacal signs 90° apart, the Bull, Lion, Scorpion, and Water Bearer. In keeping with this procedure, "al-Zarqali...made four observations near four midpoints [of zodiacal signs] between the equinoctial and solstitial points," according to the *Epitome* (III, 13, end). Copernicus also adopted this method.

A Misprint in the *Epitome*. When Rheticus arrived in Frombork, Copericus turned over to him the autograph manuscript of the *Revolutions.* After devoting nearly ten weeks to studying it, Rheticus asserted publicly that he had "mastered the first three Books."[86] In III, 16, Copernicus talked about the sun's entry into the midpoints of zodiacal signs. Rheticus knew that "it is impossible, as Ptolemy states, by means of instruments to determine with precision the times of the solstices." Rheticus also referred to "the method of intermediate positions on the ecliptic [the sun's apparent annual course through the heavens], explained by Regiomontanus in the *Epitome,* Book III, Proposition 14." Rheticus then added that, "according to Regiomontanus, al-Zarqali boasts that he made 402 observations."[87]

Actually, the *Epitome* said nothing about al-Zarqali boasting. That wrong impression was received by Rheticus, because he failed to detect a misprint in the *Epitome.* It said (III, 13, end) that al-Zarqali made "402" observations. The last character is the number 2.[88] This was a misprint for the letter "r". But Rheticus was misled by the ambiguity in the second character. He thought it was zero. This was interchangeable with the letter "O." Because of the misprint, Rheticus misread as "402" what the *Epitome* intended to be a contraction for *quattuor* or 4. The word *quattuor* appears in the very next line. This time it is spelled out in full. Somehow Rheticus did not grasp that, according to the *Epitome,* al-Zarqali "made four observations near four midpoints."

Rheticus the Confirmed Astrologer. Rheticus attended the University of Wittenberg, where he received his master's degree on 27 April 1536. Ten days earlier he had publicly defended his thesis. There he contended that Roman law did not absolutely prohibit all forms of astrological prognostication. Predictions based on physical causes, he argued, were permitted in astrology as in medicine. In his *First Report on Copernicus' Revolutions* (23 September 1539), he inserted a section asserting that "The Kingdoms of the World Change with the Motion of the Center of the Eccentric" of the sun. Throughout the rest of his life he continued to view astronomical phenomena as causes of political vicissitudes. Thus, on 1 March 1562 he wrote to a friend that he was planning a chronology from the beginning of the world "according to Copernicus' astronomy....I put my zero base in 60 B.C., at the start of the anomaly of precession [regulated by a circlet, which Rheticus called the 'Wheel of Fortune']. From this origin I move through the revolutions of the anomaly [backward] to the creation of the world and [forward] to the end of the world."[89]

Did Copernicus Believe in Astrology?

> Nothing of this theory of monarchies is mentioned by Copernicus himself, but we cannot doubt that Rheticus would not have inserted it in his account if he had not had it from his "D. Doctor Praeceptor," as he always calls him.[90]

Dreyer, an outstanding historian of astronomy, could not doubt that Rheticus' preceptor or teacher, Copernicus, developed this astrological view of political history and then had it published by Rheticus. Dreyer conceded that "nothing of this theory of monarchies is mentioned by Copernicus himself." Nor is any other astrological concept mentioned by Copernicus anywhere in his writings. This is an extraordinary aspect of Copernicus' mentality. He lived in an age when many of those in power as well as of those on the lower rungs of the social ladder believed in astrology. He did not.

Copernicus' *Euclid* and Abu l-Hasan Ali. When Copernicus attended the University of Cracow, astrology was in great vogue at that institution. He bought a copy of the *Very Famous Complete Work on the Judgment of the Stars, Which Was Written by Albohazen Haly, Son of Abenragel.* "This enormous hodge podge of astrological lore"[91] in eight books was composed by Abu l-Hasan Ali ibn Abi'r Rigial in the middle of the eleventh century. Two centuries later, it was translated from Arabic into Spanish by order of King Alfonso X of Castile, sponsor of the *Alfonsine Tables.* From Spanish, it was turned into Latin. This was first printed in Venice in 1485. Copernicus bought a copy of this first edition.

He had it bound with his copy of the first edition of Euclid's geometry (Venice, 1482). On the opening page of Euclid, he wrote his name as owner. He made annotations on Euclid.

There are annotations on Abu l-Hasan Ali also. These were once thought to be in Copernicus' handwriting, and as such provided a basis for ascribing a belief in astrology to Copernicus. The handwriting is now recognized as belonging to someone other than Copernicus.[92]

Copernicus never bought another astrological book. In his writings he never referred to astrology. Those around him did, especially Rheticus. Despite Dreyer, Copernicus did not devise the astrological theory of monarchies nor did he give it to Rheticus to be printed in his *First Report.*

Did Copernicus See Rheticus' *First Report* before it Was Printed? Did Copernicus see the *First Report* before Rheticus took it from Frombork (where there was no press) to Gdańsk to have it printed? Had Copernicus seen it, would he have allowed Rheticus to say, in a book expounding Copernicus, that al-Zarqali boasted of having made 402 observations? Copernicus himself attributed no such exaggerated number of observations to al-Zarqali. Had Copernicus noticed that number in Rheticus, would he not have questioned it, and thereby corrected Rheticus' misreading of the *Epitome*? It would appear that Copernicus did not see Rheticus' *First Report* before it was printed. He may even not have known that Rheticus included the astrological theory of monarchies in that work, the earliest public pronouncement about the Copernican astronomy.

Notes

78. Ernst Zinner, *Leben und Wirken des Joh. Müller von Königsberg genannt Regiomontanus,* 2nd ed. (Osnabrück, 1968), pp. 75, 315 #42.

79. *Dictionary of Scientific Biography,* XV (New York, 1978), 477.

80. *Proceedings of the American Philosophical Society,* 1973, vol. 117, pp. 425-426, 512.

81. A. I. Sabra, "An Eleventh-Century Refutation of Ptolemy's Planetary Theory," p. 118, in *Science and History, Studies in Honor of Edward Rosen* (Studia Copernicana XVI).

82. *Ibid.*

83. José Maria Millás Vallicrosa, *Estudios sobre Azarquiel* (Madrid/Granada, 1943-1950), p. 296/last 3 lines.

84. Ernst Zinner, "Die Tafeln von Toledo (Tabulae Toletanae)," *Osiris,* 1936, *1*:750 #34.

85. G. J. Toomer, "A Survey of the Toledan Tables," *Osiris,* 1968, *15*:45.

86. Rosen, *Three Copernican Treatises,* pp. 109-110.

87. *Ibid.,* p. 124.

88. Misread as "r" by Toomer, *Centaurus,* 1969, *14*:308/7 up.

89. Karl Heinz Burmeister, *G. J. Rhetikus* (Wiesbaden, 1967-1968), III, 162.

90. John Louis Emil Dreyer, *History of the Planetary Systems from Thales to Kepler* (Cambridge, England, 1906; reprinted as *A History of Astronomy from Thales to Kepler,* New York, 1953), p. 333.

91. As it was called by Richard Lemay, "The Late Medieval Astrological School at Cracow and the Copernican System," in *Science and History* (Studia Copernicana XVI), p. 351.

92. Photo 2, on p. 387, in Czartoryski, "The Library of Copernicus," in *Science and History* (Studia Copernicana XVI).

CHAPTER 9

The Publication of Copernicus' Astronomical Writings

The earliest of Copernicus' writings on astronomy came to be known later on as the *Commentariolus.* He did not give it this title. In fact, he did not give it any title. What is more, he withheld his name from it as author. Was he acting out of modesty or out of prudence?

He did not seek a publisher. Instead, he sent copies to professional friends. One such copy was included in an inventory of his own books and manuscripts compiled by Matthew of Miechów (1457-1523), a professor at the University of Cracow. He listed a "manuscript of six leaves expounding the theory of an author who asserts that the earth moves while the sun stands still." This length, six leaves, fits the *Commentariolus.* So does Matthew's succinct summary of the *Commentariolus'* principal innovation. Matthew finished his inventory on 1 May 1514. Therefore, the *Commentariolus,* which Copernicus did not date, must have been sent by him from Frombork to Cracow not later than the spring of 1514.

The *Commentariolus* Was Written after 15 July 1502. In a section devoted to various calculations of the length of the year, the *Commentariolus* says that "Hispalensis...determined the tropical year [measured from equinox to equinox] as 365 days, 5 hours, 49 minutes." *Hispalensis* is a Latin word signifying "from Hispalis," the ancient Roman name of the city now called Seville. For a long time scholars were unable to find an astronomer from Seville whose tropical year was $365^d5^h49^m$. Then an outstanding specialist in Copernican studies correctly identified Copernicus' "Hispalensis" with Alfonso de Cordoba Hispalensis.[93] He published a book giving the length of the year as $365\ 1/4^d - 11^m = 365^d5^h49^m$. The printing of Hispalensis' book was completed on 15 July 1502 in Venice. At that time Copernicus was a student at the nearby University of Padua. Hence, he wrote the *Commentariolus* after 15 July 1502 and before 1 May 1514.

Hispalensis Was Not King Alfonso X. The identification of Copernicus' Hispalensis with Alfonso de Cordoba Hispalensis was accepted by the academic community for half a century without demurral. Then a challenger asked: "Could *Hispalensis* in fact be a misreading of *Hispaniensis?*"[94] No manuscript of the *Commentariolus* written by Copernicus' own hand survives. Three copies are preserved. They all read *Hispalensis.* The challenger failed to specify who could have been guilty of misreading *Hispalensis* in place of *Hispaniensis.*

Unable to pin the blame on anyone for a misreading which he himself imagined, the challenger nonchalantly proceeded as though *Hispaniensis* (a Spaniard) were present in the *Commentariolus*: "*Hispaniensis*...would undoubtedly refer to good King Alfonso." An ordinary subject of King Alfonso X might be called *Hispaniensis*. But this description was never applied to Alfonso X in the *Alfonsine Tables*. Copernicus owned a copy of the second edition (Venice, 1492). It referred to its patron four times. Once it called him "king of Castile." Three times it designated him "king of the Romans and of Castile." In the inconclusive election of the Holy Roman Emperor in 1257, Alfonso was only a partially successful canididate. Had Copernicus ever referred to Alfonso X, he would never have called him *Hispaniensis*. When Copernicus referred to al-Zarqali, who was a Spaniard, he called him, not *Hispaniensis*, but *Hispanus* or *hispanus*.[95]

Al-Zarqali, King Alfonso X, and the *Alfonsine Tables* had no special connection with Seville. But the *Commentariolus'* Hispalensis did. His *Astronomical Tables for Queen Isabella* (Venice, 1503) "compute the mean motions for the meridian of Seville," present a "Table of the Rising of the Signs at Seville," offer a star catalog for Seville and Rome, and attribute the foundation of the city of Seville to no less a hero than Hercules' son, Hispalis.

The Length of the Year in the *Alfonsine Tables*. Hispalensis' $365^d5^h49^m$, according to the challenger, "is simply a rounding of the tropical year in the *Alfonsine Tables*." Where could Copernicus have found the length of the tropical year in his edition of the *Alfonsine Tables*? The challenger did not tell us.

Later versions of the *Alfonsine Tables* were modified by their editors. Half a century after Copernicus' edition, and some three decades after he wrote the *Commentariolus*, the *Alfonsine Tables* were said to subtract from 365^d6^h, not 11^m (as the *Commentariolus'* Hispalensis did) but "about 10^m44^s, which is a little more than $1/6^h$." This statement was made by a professor of astronomy at the University of Wittenberg in his edition of Peurbach's *New Theory of the Planets* (Wittenberg, 1542, sig. e4v-5r). The professor did not find $365^d5^h49^m$ as the length of the year in his edition of the *Alfonsine Tables*. But Copernicus did find $365^d5^h49^m$ in Alfonso de Cordoba Hispalensis' *Almanach perpetuum* (Venice, 1502, sig. Alv).

Did Copernicus' *Commentariolus* Use Ptolemy's *Syntaxis* as a Source? All in all, then, the correct identification of Hispalensis in the *Commentariolus* must be upheld against the mistaken objections of the challenger. He is wrong about the reading in the manuscripts, about Copernicus' term for a Spaniard, about the applicability of *Hispaniensis* to King Alfonso X, and about the length of the tropical year in Copernicus' edition of the *Alfonsine Tables*.

He is also wrong in listing (pp. 425-426), as a possible source of Copernicus' *Commentariolus,* a printed edition of Ptolemy's *Syntaxis* in Latin translation. That edition came from the press in Venice on 10 January 1515. On or before 1 May 1514, Copernicus' *Commentariolus* was entered in Matthew of Miechów's inventory, as the challenger himself indicated (p. 430). Then how could Copernicus, completing the *Commentariolus* before 1 May 1514, have consulted as a possible source a work not available until 1515?

The Planetary Loops. In the course of its eastward movement through the heavens, an outer planet advances at a constantly varying rate. Gradually the planet slows down until it seems hardly to move at all. Here it has reached its station or stationary point. It then reverses direction, proceeding in retrogradation westward with a deviation in latitude, until it reaches its second stationary point. There it resumes its regular eastward journey, and repeats this pattern of a planetary loop over and over. These planetary loops were believed by astronomers before Copernicus to be physically real. They were called the planet's second anomaly, as distinguished from its first anomaly or irregularity in the rate of the eastward motion.

The Planetary Loops Were Transformed by the *Commentariolus* Into an Optical Appearance. For the first time in the history of human thought, Copernicus' *Commentariolus* declared that "this second anomaly happens by reason of the motion, not of the planet, but of the earth." In Copernicus' astronomy, the earth revolves once a year around the sun in a path which he called the Grand Orb *(orbis magnus).* This is completely enclosed within the orbit of an outer planet, which is called "outer" for that very reason.

Imagine these three bodies (sun, earth, outer planet) in a straight line. The sun is stationary in the center, with the earth and outer planet to either side of it. These two bodies revolve around the sun in the same direction, eastward. Hence, to an observer on the earth, watching the planet against the background of the stars, the planet seems to be advancing. But it does so with diminishing speed, because its angular velocity is slower than the earth's. When the boundary condition is reached, although the planet continues to move eastward at a rate slower than the earth's, it seems to the terrestrial observer to be standing still. Soon, with the earth moving faster than the outer planet, against the background of the stars the outer planet seems to be moving backward or retrogressing. At the next boundary condition, the outer planet again seems stationary. But it soon resumes its eastward advance, at first slowly, then more rapidly.

As the *Commentariolus* explained, the outer planet never stops and never changes direction. It only seems to do so. The reason for this appearance is that it is watched from a moving platform, the earth. *(See Reading No. 47.)*

This transformation of the planetary loops from a physical reality to an optical appearance was an invincible argument for the validity of the astronomy of Copernicus. His analysis has been repeated for centuries in elementary handbooks of science all over the world.

The *Commentariolus* Placed an Outer Planet's Stationary Point Where Pliny the Elder Did. In the *Commentariolus'* discussion of the outer planets, Copernicus stated where the earth and the planet would generally be when the planet was seen at its stationary point. As the earth and the outer planet each moved in its own orbit around the stationary sun, about 120° would separate the sun from the planet as seen from the earth:

> When the line of sight is moving in the direction opposite to the planet's and at an equal rate, the planet seems to stand still because the opposite motions neutralize each other in this way. This generally happens when the angle at the earth between the sun and the planet is about 120°.

This imprecise approximation was taken by Copernicus from the *Natural History* of Pliny the Elder (c. 23-79):

> In the trine aspect, that is, at 120° from the sun, the three outer planets have their morning stations, which are called the first stations...and again at 120°, approaching from the other direction, they have their evening stations, which are called the second stations (II, 59).

Copernicus himself did not possess his own copy of Pliny's *Natural History,* a very expensive book. The Varmia Chapter's copy of the Rome 1473 edition is still preserved. It is annotated by several hands. Whether Copernicus wrote any of these annotations has not yet been decided. But in a copy of the Venice 1487 edition he transcribed a passage from Cicero, which is unmistakably in his handwriting.[96]

When Did Copernicus Find Out about Apollonius of Perga's Geometrical Determination of the Planetary Stationary Points? Pliny was an extremely energetic and diligent researcher, who consulted nearly 500 previous authors. Yet he was not aware that about two and a half centuries earlier the great mathematician Apollonius of Perga had shown how to find a planet's stationary points geometrically. This demonstration was contained in a treatise by Apollonius that as an entity was lost before Pliny's time. Yet, less than a century later, Apollonius' theorems concerning the stationary points were restated in Ptolemy's *Syntaxis,* XII, 1. This Greek discussion was reproduced in Latin about 1300 years later by Regiomontanus in the *Epitome,* XII, 1-2. The *Epitome,* V, 22, pointing out the defect in Ptolemy's lunar theory, was used by Copernicus in the *Commentariolus.* But at that time *Epitome,* XII, 1-2, was unfamiliar to him. For had he then known *Epitome,* XII, 1-2, he would have discarded from the *Commentariolus* Pliny's primitive approximation of 120° between the sun and the stationary planet, as seen from the earth.

Nearly three decades later, in writing *Revolutions,* Book V, Copernicus was so familiar with Apollonius' theorems that he found them defective. *(See Reading No. 48.)*

An Interesting Technical Term in *Revolutions,* V, 35. After reviewing Apollonius' treatment of the planetary stations in *Revolutions,* V, 35, Copernicus remarked that "Apollonius adduces a certain auxiliary theorem." Copernicus calls it *lemmation,*[97] a small lemma or proposition. This Greek word does not occur in the Latin sources consulted by Copernicus. He had previously used *demonstrata,* a Latin word, in *Revolutions,* V, 3,[98] but he noticed *lemmation* in the Greek text of Ptolemy's *Syntaxis.* A copy of its first edition (Basel, 1538) was presented to him in the following year by Rheticus on his arrival in Frombork.

The Displacement of V, 35, in the Autograph Manuscript of the *Revolutions.* Copernicus started to write *Revolutions,* V, 35, in his autograph on folio 188 recto. There he inscribed the title "On the Stations and Retrogradations of the Five Planets." Below it, however, he placed the symbol of the zodiacal sign of the Balance in an enlarged form. He left the rest of folio 188r blank. On the verso of folio 188 he began Book VI, which he continued to folio 197 recto. Then on folio 197 verso he repeated the symbol for the zodiacal sign of the Balance at the top, and below it the chapter title "On the Stations and Retrogradations of the Five Planets." Chapters 35 and 36, the last two chapters of Book V, followed.

Order and Disorder in the Autograph Manuscript of the *Revolutions.* In his autograph manuscript of the *Revolutions* Copernicus entered numerous corrections, deletions and additions. But the folios followed one another in proper sequence until Book V, 30. At the bottom of folio 182 recto, Copernicus placed the symbol for the zodiacal sign of the Fishes, signifying an interruption. He repeated this symbol at the top of folio 195 recto, where he resumed Book V, 30. Somewhat similarly, he began Book V, 35, on folio 188 recto, and then started Book V, 35, all over again on folio 197 verso.

These folio numbers were not written in by Copernicus himself. They were inscribed for a later proprietor of his autograph manuscript. As it left his hands, the closing chapters of Book V and all of Book VI must have looked bewilderingly disorderly to anyone not thoroughly familiar with its contents. No wonder a fair copy was made for the convenience of the typesetters.

When Did Copernicus Finish Writing the *Revolutions?* The books presented to Copernicus by Rheticus on his arrival in Frombork produced a minor upheaval in the autograph manuscript of the *Revolutions.* Regiomontanus' treatise *On All Kinds of Triangles* (Nuremberg, 1533) persuaded Copernicus to revise his trigonometrical section. Thus, he cancelled his original draft of a theorem, which he expanded. The numbering of the theorems was revised, more than once.

In his *First Report,* when Rheticus began to discuss the planets, he indicated that Copernicus "has in large measure already accomplished" his labors. Rheticus thereby implied that Copernicus' labors on planetary theory had not yet been completed. This implication holds true for the period before 23 September 1539, when Rheticus finished writing the *First Report.* On 15 April 1541 a Wittenberg friend relayed the information that "Professor [George] Joachim [Rheticus] has written from Prussia that he is waiting for his teacher [Copernicus] to finish his work,"[99] the *Revolutions.* As late as 2 June 1541 Rheticus still reported to the same friend that his "teacher feels quite well and is writing a great deal."[100]

In the meantime Rheticus was busy composing his own *German Topography (Chorographia tewsch).* In dedicating it to Duke Albert of Prussia in August 1541, Rheticus remarked that through Copernicus' praiseworthy work we shall have an accurate computation of time and account of the heavenly motions. Then on 29 August 1541 Rheticus presented to the duke an instrument for determining the hour of daybreak throughout the year. In that same document Rheticus asked the duke to request the Elector of Saxony and the University of Wittenberg to permit Rheticus to publish Copernicus' *Revolutions (See Reading No. 49.)* The duke's secretary promptly wrote to both the elector and the university on 1 September 1541 a highly eulogistic letter. But the secretary made the unfortunate mistake of misidentifying the author of the book to be printed as Rheticus himself, instead of Rheticus' teacher, Copernicus. *(See Reading No. 50.)*

Rheticus was back in Wittenberg in time to be elected dean of the faculty for the winter semester beginning 18 October 1541. He had brought with him for the printer a fair copy of Copernicus' *Revolutions.* That work must have been finished, then, before Rheticus left Frombork in September 1541.

Copernicus' *Sides and Angles of Triangles* Was Published in Wittenberg. From Copernicus' *Revolutions,* Rheticus extracted the section on trigonometry and published it in Wittenberg early in 1542 as Copernicus' *Sides and Angles of Triangles.* This differs somewhat from the corresponding material in the *Revolutions.* For example, the *Revolutions'* table of sines is based on a radius of 100,000, with the central angle increasing by intervals of 10'. By contrast, the 1542 table's radius is 10,000,000, or two places longer, while the central angle increases by intervals of only 1'. The complementary angle is indicated at the foot of the columns and at the right-hand side of the page. This is the earliest table to give the cosine directly (that term is not used, just as "sine" is avoided). Although this table was undoubtedly drawn up by Rheticus, he did not ascribe it to himself, presumably out of modesty. Nor did he attribute it to Copernicus.

In Wittenberg, the mightiest fortress of anti-Catholic Lutheranism, Rheticus identified the author of *Sides and Angles of Triangles* geographically with his birthplace Toruń, not confessionally with the Roman Catholic Cathedral Chapter of Frombork. In the preface Rheticus explained that Copernicus "wrote most learnedly about triangles while he was hard at work elucidating Ptolemy and expounding the theory of the [heavenly] motions." "Elucidating" *(illustrando)* Ptolemy, not contradicting him, was what Rheticus depicted Copernicus as doing when the *Sides and Angles of Triangles* was published in Wittenberg.

No such subterfuge could be invoked when it came to publishing the *Revolutions* with its undisguised contradiction of Ptolemy. No wonder Rheticus told the duke of Prussia's secretary that he wanted permission to "betake himself for a time, without interruption of his professor's wages, for the sake of carrying out his intended work in the place where he decided to have the book printed."

Copernicus' *Revolutions* Was Published in Nuremberg. The duke's secretary understood that the place of publication was not going to be Wittenberg. What Rheticus did not tell him (and the duke) was that the place of publication would be Nuremberg, and the publisher-printer, Johannes Petreius (1497-1550). In August 1540 Petreius dedicated one of his publications to Rheticus. In this dedication Petreius told his readers that Rheticus had gone to a distant part of Europe. There he published a brillant exposition of an unconventional astronomer's theory. Petreius then added: "I deem it a marvelous treasure if his observations are communicated to us at your instigation at some future time, as we hope will come to pass." Rheticus had made this arrangement with Petreius in Nuremberg because he sensed how strong the opposition to Copernicanism was in Wittenberg.

Martin Luther's Rejection of Copernicanism. The dominant figure in Wittenberg was Martin Luther (1483-1546). Not long after Rheticus began his journey to visit Copernicus, on 4 June 1539 the conversation among the guests in Luther's dining room turned to the innovative astronomer. It was a common practice among the guests to write down what Luther said during his meals. Twenty years after his death, the earliest edition of his *Table Talk (Tischreden)* appeared in 1566. Here Luther called Copernicus a *"fool" (Narr),*[101] a characterization repeated in later editions of the *Table Talk.* But this condemnation, in keeping with Luther's customary vehemence, was softened in an alternative version. *(See Reading No. 51.)*

Relativity of Motion in Luther's *Table Talk*. In Luther's *Table Talk,* both versions of the conversation about Copernicus are in agreement that he wanted to prove the moon does not move. In reality Copernicus wanted to prove no such thing. He was discussed in Luther's dining room while being visited by Rheticus. Neither Rheticus' *First Report* nor Copernicus' *Revolutions* had been printed as yet. Either work would have instantly dispelled the misconception that Copernicanism denied the moon's motion.

Yet the principle of relative motion was understood in Luther's dining room on 4 June 1539. Copernicus' application of this principle, according to Luther, turned astronomy upside down. According to Copernicus, it turned astronomy right side up. *(See Reading No. 52.)*

Philip Melanchthon's Rejection of Copernicanism. Philip Melanchthon (1497-1560), Luther's principal lieutenant, shared his leader's contempt for Copernicus. When Rheticus returned to Wittenberg from Frombork after spending more than two years with Copernicus, on 16 October 1541 Melanchthon wrote to a friend:

> Certain people think it is a wonderful achievement to elaborate such an absurd thing, like that Polish astronomer who moves the earth and holds the sun stationary. Certainly prudent governments ought to restrain the smart alecks' insolence.[102]

No *Revolutions* in Wittenberg. Melanchthon was the dominant personality in the University of Wittenberg. His attitude that "prudent governments ought to restrain the smart alecks' insolence" was a signal that Copernicus' *Revolutions* could not be printed in Wittenberg. Melanchthon's hostility was not directed against Rheticus personally. He was allowed to publish in Wittenberg Copernicus' *Sides and Angles of Triangles,* a purely technical book. But Copernicus' *Revolutions* entailed such devastating theological implications that Rheticus could not hope to obtain permission for it to be printed in the innermost citadel of Lutheran orthodoxy.

Copernicus' Dedication of the *Revolutions*. At the end of the winter semester on 1 May 1542, Rheticus took his last leave of absence from the University of Wittenberg, to which he never returned. He headed directly to Petreius' workshop in Nuremberg, where the first two signatures of Copernicus' *Revolutions* were printed by the end of May. In June, Copernicus composed his magnificent Dedication of the *Revolutions* to the reigning pope, Paul III, whom he asked for protection against calumnious attacks *(See Reading No. 53.)*

Was the *Revolutions* Mentioned in the *Commentariolus?* In the *Commentariolus,* Copernicus said: "I have thought it well, for the sake of brevity, to omit from this sketch mathematical demonstrations, reserving these for my larger work *(maiori volumini).*"[103] The only "larger work" ever written by Copernicus is his *Revolutions,* in which he inserted the mathematical demonstrations omitted from his *Commentariolus.* Evidently he was already planning the *Revolutions,* or had already begun to write it, when he made this reference to it in his *Commentariolus.*

A challenger has asked: "What is the 'larger book' he [Copernicus] refers to? There is no reason to believe it was to be anything like the *Revolutions*....I believe that the sort of book Copernicus was contemplating when he wrote the *Commentariolus* would have consisted of geometrical demonstrations of the equivalence of Ptolemy's and his own models."[104]Nothing was ever farther from Copernicus' mind than to equate his own cosmology with Ptolemy's. Copernicus correctly saw that the earth was a heavenly body, a planet, in motion. Ptolemy's earth was a non-heavenly body at rest.

The *Commentariolus*' Reference to the *Revolutions* Is Confirmed by the Dedication of the *Revolutions*. In the Dedication of the *Revolutions*, Copernicus explained that he debated with himself "for a long time whether to publish the volume which [he] wrote to prove the earth's motion." *(See Reading No. 54.)* He recalled that the ancient Roman author Horace, in his *Art of Poetry* (lines 388-389), advised budding writers not to publish their work as soon as it was finished, but to hold it back "until the ninth year thereafter." Accepting Horace's advice, Copernicus quadrupled it. As he said in the Dedication of the *Revolutions*, a friend "urgently requested me to publish this volume and finally permit it to appear after being buried among my papers and lying concealed not merely until the ninth year but by now the fourth period of nine years."[105]

A fourth period of nine years *(quartum novennium)* begins after 27 years. Copernicus wrote the Dedication in 1542. Subtracting 27 from 1542 leaves 1515 as the latest year in which "this volume" began to lie concealed among Copernicus' papers. The *Commentariolus* was written before 1 May 1514. Hence, the Dedication of the *Revolutions* confirms the *Commentariolus*' reference to the *Revolutions*: Copernicus began to write the *Revolutions* not later than 1515.

The challenger soon abandoned his effort to identify the *Commentariolus*' "larger work" as some (nonexistent) equivalent of Ptolemy. Instead, he said: "What kind of larger book Copernicus had in mind when he wrote this [reference to the *Revolutions* in his *Commentariolus*] is by no means clear."

Copernicus Did Not Begin to Write the *Revolutions* in 1530 or Thereafter. The challenger was guilty of falsifying the evidence when he professed to be quoting from the Dedication of the *Revolutions*. There Copernicus said he concealed "this volume" *(librum hunc)*, the *Revolutions*. Instead, the challenger pretended that Copernicus said he "concealed his theories." But Copernicus did not conceal his theories when he circulated the *Commentariolus* before 1 May 1514. He did start to conceal "this volume," the *Revolutions*, by 1515.

The challenger also declared that it "is by no means clear what kind of larger book Copernicus had in mind when he wrote this" unambiguous reference to the *Revolutions* in his *Commentariolus*. Why ask "what kind of larger book"? Copernicus had only one kind of larger book in mind, the *Revolutions*. He devoted nearly three decades of his life to writing it.

As the third step in his misrepresentation of the pertinent facts, the challenger asserted: "In all probability the *Revolutions* was entirely a work of the 1530's, and I can see no evidence for giving any part of it an earlier date."[106] In *Revolutions*, I, 11, Copernicus remarked that, in addition to the eight celestial spheres accepted since ancient times, " some people...adopted a ninth surmounting sphere. This having proved inadequate, more recent writers now add on a tenth sphere." But in *Revolutions*, III, 1, he had still more recent information: "Some people excogitated a ninth sphere, and others a tenth...An eleventh sphere too has already begun to emerge into the light of day." When did Copernicus, in III, I, learn of an eleventh sphere, about which he knew nothing in I, 11?

Copernicus' *Letter Against Werner*. Johann Werner (1468-1522) published a collection of mathematical studies (Nuremberg, 1522). These included (at signatures k lr - v 4r) an essay on "The Motion of the Eighth Sphere," or sphere of the stars. A copy of this essay was sent to Copernicus to elicit his opinion. He expressed this in his *Letter against Werner*, dated 3 June 1524.

A commentator on Peurbach referred to those who "conceived two spheres above the eighth [sphere], as was done by Peurbach...in imitation of the opinion of the *Alfonsine [Tables]*....Werner added an extra sphere on top of the spheres of the *Alfonsine [Tables]*."[107] Copernicus learned about an eleventh sphere from Werner.

This eleventh sphere was unknown to Copernicus when he wrote *Revolutions*, I, 11. But in *Revolutions*, III, 1, he said: "An eleventh sphere too has already begun to emerge." He found out about this emergence between 1522 and 1524. Hence, he wrote *Revolutions*, I, 11, before 3 June 1524, and *Revolutions*, III, 1, not long thereafter.

Aristyllus, Not Aristarchus. The 1515 Latin translation of Ptolemy's *Syntaxis* said (folio 73r/22 up -20 up) that Hipparchus "found that there were very few extant observations of the fixed stars before him, and he found that only the observations of Arsatilis and Timocharis were recorded." "Arsatilis," as Copernicus recognized, was an Arabic distortion of the name of a Greek astronomer. At first Copernicus misidentified "Arsatilis" with Aristarchus. He did so in his *Letter against Werner* as well as in the autograph manuscript of the *Revolutions*, folio 78 verso, line 2. Later he found out that the correct identification was Aristyllus, which he entered in the right margin of folio 78v. Hence he wrote Aristarchus on folio 78v as part of *Revolutions*, III, 6, before 3 June 1524, when he made the same mistake in his *Letter against Werner*, dated 3 June 1524.

Why add any further proofs that Copernicus wrote part of the *Revolutions* before 1530, as against the challenger's refusal to give "any part of it an earlier date"? Copernicus began to write the *Revolutions* about 1515, and finished it in 1542.

Notes

93. Ludwik Antoni Birkenmajer, *Stromata Copernicana* (Cracow, 1924), p. 353.

94. *Proceedings of the American Philosophical Society,* 1973, *117*:452.

95. *Copernicus Complete Works,* I, fol. 73r/23, 79r/23, 97r/11 up.

96. Facsimile in L. A. Birkenmajer, *Mikołaj Kopernik,* facing p. 567.

97. *Copernicus Complete Works,* I, folio 197v/line 2 up.

98. *Ibid.,* fol. 149v/last line.

99. *Corpus reformatorum,* IV (Halle, 1837), #2194; reprinted, New York, 1963.

100. Burmeister, *Rhetikus,* III, 271/5-6.

101. *Tischreden oder Colloquia Doct. Mart. Luthers,* ed. Joannes Aurifaber (Eisleben, 1566), folio 580 recto/25-34.

102. *Corpus reformatorum,* IV, column 679, #2391.

103. Rosen, *Three Copernican Treatises,* p. 59.

104. *Proceedings of the American Philosophical Society,* 1973, *117*:439.

105. Copernicus, *On the Revolutions,* p. 3/35-37.

106. *Journal for the History of Astronomy,* 1974, *5:* 188, 194.

107. E. O. Schreckenfuchs, *Commentaria in novas theoricas planetarum Georgii Purbachii* (Basel, 1566), pp. 388-389.

CHAPTER 10

Copernicus and the Scientific Revolution

Copernicus began the Dedication of the *Revolutions* by explaining that his anticipation of the hostile reaction he would encounter almost caused him to abandon the work. *(See Reading No. 54.)* His fear was not overcome, even when he received a letter from Nicholas Cardinal Schönberg (1472-1537), offering to have his writings copied at the cardinal's expense and sent to Rome. *(See Reading No. 55.)* The cardinal's letter, dated 1 November 1536, was carefully filed by Copernicus. But he did not do what the cardinal asked. He did not communicate his discovery to scholars. He did not send his writings to the cardinal. He did not allow everything to be copied in his quarters at the expense of the cardinal, who died on 9 September 1537.

Bernardino Baldi's Biography of Copernicus. The relation between Copernicus and Cardinal Schönberg was completely distorted in the earliest surviving substantial biography of Copernicus. This was completed on 7 October 1588 by Bernardino Baldi (1553-1617). Relying mainly on his fertile imagination, Baldi stated:

> Schönberg had Copernicus' work; recognized its perfection and excellence; showed it to the pope, by whose judgment it was approved. The said Cardinal [Schönberg] addressed himself to Copernicus to ask him for many reasons to be willing to publish it....Copernicus dedicated it to Pope Paul III, by whose judgment, as has been said, it had been approved. What reward Copernicus obtained for it and what happened in the said affair, I would not know.[108]

Did Pope Paul III Approve of Copernicus' *Revolutions?* Despite Baldi's inventiveness, Cardinal Schönberg, having no manuscript copy of Copernicus' *Revolutions,* could not have shown it to Pope Paul III, who was therefore in no position to approve or disapprove of the *Revolutions* before it was printed. But after it was printed, it was submitted to the pope's personal theologian, who bore the title Master of the Sacred and Apostolic Palace. As his Master of the Sacred and Apostolic Palace, Pope Paul III in July 1542 appointed a friar of the Dominican Order, Bartolomeo Spina of Pisa.

> Because of Spina's outstandingly gifted mind, the pope thought most highly of him. Accordingly he relied on his advice in the difficult matters of faith which arose at that time, and the pope wanted Spina to be one of the five selected men whom the pope constituted at Rome to judge the questions raised at the Council of Trent.[109]

The issues discussed at the Council of Trent were so momentous that Spina did not carry out to the end his review of the *Revolutions.* But a fellow Dominican and lifelong friend of Spina reported that

> The Master of the Sacred and Apostolic Palace had planned to condemn his [Copernicus'] book. But, prevented at first by illness, then by death, he could not carry out this [plan]. This I took care to accomplish afterwards in this little work for the purpose of safeguarding the truth to the general advantage of Holy Church.

This was how Giovanni Maria Tolosani (c. 1471-1549) concluded his *Heaven and the Elements,* the fourth appendix to his treatise *On the Truth of Holy Scripture.* Baldi and others who spoke of Pope Paul III's approval of Copernicus' *Revolutions* never could provide documentary proof of such approval. It has been relegated to the trash basket of historical myths by the recent publication, for the first time, of Tolosani's account of Spina's plan to condemn the *Revolutions.* What Spina planned, but did not live long enough to do, was done by Tolosani. *(See Reading No. 56.)*

The Beginning of the Rheticus-Osiander Correspondence. While Rheticus was visiting Copernicus in Frombork, he wrote a letter about his host to Andreas Osiander (1498-1552) in Nuremberg. A prominent Lutheran preacher, Osiander was also keenly interested in the mathematical sciences. For that reason Rheticus offered his friendship to Osiander. The latter replied on 13 March 1540, telling Rheticus how highly he valued Copernicus' intellect. For the time being, he did not dare to write to Copernicus. But he did ask Rheticus to let Copernicus know that his friendship would be appreciated by Osiander too. *(See Reading No. 57.)*

Rheticus Sends Presentation Copies of his *First Report* to Osiander. When the printing of Rheticus' *First Report* was completed in Gdańsk in 1540, he sent presentation copies to several friends, including Osiander. The latter's acknowledgement that he had received a number of copies was expressed in a letter, of which only the opening lines are preserved. They were discovered recently, and published for the first time. *(See Reading No. 58.)*

The Copernicus-Osiander Correspondence. When Osiander read Rheticus' *First Report,* he overcame his reluctance to write to Copernicus. He now dared to do what he previously had not dared to do on 13 March 1540. Osiander challenged Copernicus' attitude toward astronomical hypotheses in a letter that has not been preserved. Copernicus' reply, dated 1 July 1540, has likewise been lost. It took eight months to reach Osiander, who received it in March 1541. On 20 April 1541 Osiander replied to Copernicus. Only part of his response survives. *(See Reading No. 59.)*

Osiander's Companion Letter to Rheticus. By the same courier who took Osiander's letter to Copernicus on 20 April 1541, the Lutheran minister sent a companion letter to Rheticus. In it he set forth the strategy underlying what he wanted Copernicus to say in an introduction to the *Revolutions*. Only a part of Osiander's letter to Rheticus of 20 April 1541 has survived. *(See Reading No. 60.)* While the Copernicus-Rheticus-Osiander correspondence was still intact, in the hands of that master strategist, Johannes Kepler (1571-1630), his judgment was: "Strengthened by a stoical firmness of mind, Copernicus believed that he should publish his convictions openly, even though this science should be damaged."[110] If Copernicus' astronomical hypotheses yielded computations conforming exactly with the observed phenomena, according to Osiander's recommendation, he need not concern himself with the question whether his hypotheses were true or not. In other words, a correct conclusion could follow from a false premise; a fictitious hypothesis could lead to an accurate result. This fictionalist outlook was not acceptable to Copernicus. To him it made all the difference in the world whether his hypothesis that the earth moves was true or not. He believed that it was true. He must still tell the world that it is true. Those who disliked this truth might damage the science of astronomy because it discovers such disturbing facts. But, Copernicus felt, the truth must be told, anyway.

Rheticus Transfers to the University of Leipzig. While Rheticus was supervising the printing of Copernicus' *Revolutions* in Nuremberg, the University of Leipzig was undergoing a drastic reorganization. The presiding personality was Joachim Camerarius Sr. (1500-1574), a vigorous humanist receptive to fresh ideas. After some indecision between returning to the University of Wittenberg for the winter semester of 1542-1543 or accepting an offer from the University of Leipzig, Rheticus finally adopted the latter course. As far as Copernicus' *Revolutions* was concerned, Rheticus made a happy choice. After Copernicus' masterpiece was published in 1543, Melanchthon's attitude toward it softened somewhat. But Camerarius was enthusiastic about it. He wrote an imaginary dialog between a guest and an unnamed scholar (undoubtedly himself) concerning Copernicus' *Revolutions*. Camerarius composed the dialog in ancient Greek. This was later translated into Latin by Kepler on 22 December 1598. He wrote out the translation with his own hand on the flyleaf of his personal copy of the *Revolutions*. This has been preserved in the library of the University of Leipzig. In 1965 Edition Leipzig published a facsimile of this historically important copy for distribution in the socialist countries, while Johnson Reprint Corporation distributed it elsewhere.

Osiander's Fraudulent "Address to the Reader" of Copernicus' *Revolutions*. Osiander suggested to Copernicus that in an introduction to the *Revolutions* he should say that in expounding his hypotheses, he was making no claim that they are true. Copernicus rebuffed Osiander's suggestion, as we know from Kepler's summary of the letter written by Copernicus to Osiander - a letter which Kepler had and we do not. However, Osiander did not give up. When Rheticus left Nuremberg for Leipzig in October 1542, Osiander became editor of the *Revolutions,* Books V-VI. After the complete text was printed, it was the turn of the front matter. The handwritten sheets submitted to Petreius by Osiander contained an "Address to the Reader." It was written by Osiander, but he withheld his name as its author. It flatly contradicted Copernicus' *Revolutions.* Petreius, however, did not scrutinize the contents, leaving such supervision to the editor. As a result, the *Revolutions* issued from the press in a patently self-contradictory form. *(See Reading No. 61.)*

The Protest against the Interpolated "Address to the Reader" in the *Revolutions.* When copies of the self-contradictory *Revolutions* reached Rheticus in Leipzig, he flew into a rage. He sent two copies to Tiedemann Giese, the best friend of Copernicus. The astronomer, having died on 24 May 1543, could no longer defend himself. Hence, Giese drew up a letter of protest. *(See Reading No. 34.)* Giese asked Rheticus to transmit the protest to the City Council of Nuremberg. They in turn referred it to Petreius. His answer was just as tart as the protest. Neither document has survived. But the City Council's action on Wednesday, 29 August 1543, as recorded by its secretary, Jerome Baumgartner, is preserved. *(See Reading No. 62.)*

Osiander's Conversation with Philip Apian. Osiander was compelled to leave Nuremberg for religious reasons about 18 November 1548. In his search for a post elsewhere, he may have returned to Ingolstadt, where on 9 July 1515 he had been admitted to the local university. If he visited Ingolstadt in 1548, he would undoubtedly have consulted its famous professor of mathematics. This professor's son, Philip Apian (1531-1589), had a historically important conversation with Osiander - a conversation that may have taken place in Ingolstadt in November or December 1548. For on 27 January 1549 Osiander arrived in Koenigsberg, where the recently founded university appointed him professor of theology. He remained in Koenigsberg until his death on 17 October 1552.

Osiander's Private Admission that he Interpolated the "Address to the Reader" in the *Revolutions.* On 1 March 1570 Apian became professor of mathematics at the University of Tübingen. A student there purchased a copy of the first edition of the *Revolutions* on 6 July 1570. This copy is now preserved in the municipal library of Schaffhausen, Switzerland. In the front matter, at the top of folio 2 recto, the purchaser, Michael Maestlin (1550-1631), wrote a passage ending with the statement that he had been told by Apian that "Osiander openly admitted to him that he had added this [Address to the Reader] as his own idea." *(See Reading No. 63.)*

Rheticus and Ramus. By suppressing his name as author of the "Address to the Reader," Osiander misled many readers of the *Revolutions.* Even so astute a scholar as Peter Ramus (Pierre de la Ramée, 1515-1572) supposed that Rheticus wrote the "Address." On 25 August 1563 Ramus sent Rheticus a letter praising his *Canon of the Doctrine of Triangles* (Leipzig, 1551). This brief work was Rheticus' most influential contribution to the mathematical sciences. It was the first table to give all six trigonometrical functions. It defined them as ratios of the sides of a right triangle, and related them directly to the angles. By equating the functions of angles greater than 45° with the corresponding cofunctions of the complementary angles smaller than 45°, Rheticus reduced the length of his table by half. This was what impressed Ramus, who aimed to simplify mathematics and astronomy. *(See Reading No. 64.)*

Ramus' Appeal to the German Astronomers. At the end of his letter to Rheticus, Ramus recommended that any reply should be sent through Joachim Camerarius, a devoted friend of Rheticus. The latter informed another friend on 12 April 1564: "France too invites me. But I have not yet decided what I shall do."[111] In 1568, however, Rheticus sent Ramus a long list of his projects, stating bluntly: "I am founding a German astronomy for my Germans."[112] Meanwhile Ramus, an ardent Protestant, was harassed by Roman Catholic adversaries in France. He visited the Protestant universities in Germany, and was deeply impressed by what he saw. Hence in his *Scholae mathematicae* (Paris/Basel, 1567-1569) he made a public appeal to the German astronomers similar to his previous private appeal to Rheticus in 1563. To any German philosopher-scientist who would construct an astronomy without hypotheses, Ramus promised a royal lectureship in Paris, his own, if necessary. Little did he then know that in the early hours of 24 August 1572 he would be one of the many victims in the unforgettable massacre on St. Bartholomew's Day. *(See Reading No. 65.)*

Kepler Would Have Claimed Ramus' Lectureship. Ramus' private appeal to Rheticus did not elicit an astronomy without hypotheses, nor did his public call to German astronomers in general produce any immediate response. But four decades later, in 1609 on the verso of the title page of Kepler's *New Astronomy* Ramus' call was reprinted and answered. Had Ramus still been alive and holding the royal lectureship, Kepler would have claimed it for himself on the basis of his *New Astronomy,* in which he demonstrated for the first time in history that the orbit of a planet is an ellipse. Kepler's answer also publicly disclosed for the first time that the author of the "Address to the Reader" in Copernicus' *Revolutions* was Osiander. *(See Reading No. 66.)*

Why Did Copernicus Revive the Concept of the Earth as a Moving Planet? In reviving the ancient concept of the earth as a moving planet, Copernicus had to take the earth out of the center of the universe, where it had been stationed for the longest time by the greatest authorities. Copernicus does not give us a detailed report of the steps in the reasoning which led him to his reconstruction of the traditional astronomy. In the absence of such a detailed report, a challenger has called attention to Copernicus' use of Regiomontanus' *Epitome*, Book XII, Propositions 1-2:

> The eccentric model for the second anomaly [is] mentioned briefly by Ptolemy in *Syntaxis*, XII, 1...; it is this alternate model that leads directly to the heliocentric theory....Ptolemy had said that the eccentric representation of the second anomaly was usable only for the superior planets, but in *Epitome*, XII, 2, Regiomontanus describes an equivalent eccentric model for the inferior planets....I believe that Copernicus arrived at the heliocentric theory after a careful investigation of these two propositions in Book XII of the *Epitome*.
>
> ...Copernicus was investigating an alternative eccentric model of the second anomaly that was described in detail by Regiomontanus in Book XII of the *Epitome*....The model leads directly to the heliocentric theory, although its two forms for the superior and inferior planets lead respectively to the Tychonic and Copernican theories. Copernicus' derivation of his theory rests upon the eccentric model of the second anomaly and therefore upon these two propositions in the *Epitome*. In this way Regiomontanus provided the foundation of Copernicus' great discovery. It is even possible that, had Regiomontanus not written his detailed description of the eccentric model, Copernicus would never have developed the heliocentric theory. Regiomontanus...was, through these two propositions, virtually handling it [the heliocentric theory] to any taker.[113]

If "to any taker," why not to Regiomontanus himself? If "the eccentric model for the second anomaly mentioned briefly by Ptolemy...leads directly to the heliocentric theory," why did it not lead Ptolemy directly to that theory? Why did it not lead Ptolemy's predecessor Apollonius to the heliocentric theory? Why did Tycho Brahe reject Copernicus' heliocentrism?

Epitome, XII, Propositions 1-2, treat the stationary points of the planets. They present neither Copernicus' nor Tycho Brahe's alternative to Ptolemy's cosmology. They first came to Copernicus' attention after he had circulated his earlier version of the heliocentric theory in the *Commentariolus*. There he used Pliny's *Natural History* for the planetary stationary points. Only much later, in the *Revolutions*, was he aware of Apollonius' theorems, as presented in Ptolemy. Copernicus was not led to the heliocentric theory by Regiomontanus' *Epitome*, XII, 1-2, any more than Regiomontanus was. *(See Reading No. 48.)*

What Tycho Brahe Thought of Copernicus. Tycho Brahe (1546-1601), the greatest astronomer of the second half of the sixteenth century, refused to follow Copernicus in making the earth a moving planet. In Brahe's native country, Denmark, the religious party in ascendancy viewed the belief in the moving earth as being in conflict with the Bible. Brahe avoided a quarrel with the religious authorities, since he was unavoidably involved in more quarrels than he could handle. Nevertheless, when he delivered a lecture to the entire faculty and student body of the University of Copenhagen early in September 1574, he declared:

> In our time Nicholas Copernicus may not undeservedly be called a second Ptolemy. Through observations made by himself he discovered certain gaps in Ptolemy, and he concluded that the hypotheses formulated by Ptolemy admit something unsuitable in violation of the axioms of mathematics. Moreover, he found the Alfonsine computations in disagreement with the motions of the heavens. Therefore, with wonderful intellectual acumen he formulated different hypotheses. He restored the science of the heavenly motions in such a way that nobody before him reasoned more accurately about the movements of the heavenly bodies.[114]

Brahe was born only three years after the death of Copernicus and the publication of the *Revolutions.* Like Copernicus, Brahe wrote mainly in Latin. He knew that language thoroughly, unlike some people who presume to write about Copernicus nowadays. Brahe studied the *Revolutions* in intimate detail. He was aware that Copernicus' disagreement with the astronomy accepted in his time rested on two foundations. First, the *Alfonsine Tables* were not in accord with what was observed in the heavens. Secondly, the Ptolemaic system clashed with the theoretical requirements of astronomy, as then understood.

Copernicus' Own Explanation of his Reason for Regarding the Earth as a Planet. In order to account for nonuniform planetary motion, Ptolemy and the Ptolemaists permitted a circular motion to be nonuniform with respect to its own center. This Ptolemaic arrangement in the theory of the moon and the planets was rejected by Copernicus. It was this Ptolemaic departure from strict adherence to fundamental principles that gave Copernicus "the occasion to consider the motion of the earth." *(See Reading No. 67.)*

Political Revolutions and Scientific Revolutions. The term "revolutions" appeared for the first time in the title of a book on astronomy in Copernicus' masterpiece. What he meant by a "revolution" was the completion of a circular (or spherical) movement through 360°. A historian of astronomy, who later played a prominent part in the French Revolution, declared that Copernicus had effected an astronomical revolution. Copernicus' innovative pronouncements were subsequently matched and surpassed in astronomy and other sciences. Designations such as the Newtonian revolution, the chemical revolution, the Einsteinian revolution proliferated. Outside science, the Commercial Revolution, the Industrial Revolution, the Agricultural Revolution, and similar rubrics abounded.

In the history of modern nations, an effort was made to find common features in four major political revolutions. This model was followed by Thomas S. Kuhn's *Structure of Scientific Revolutions* (2nd ed., Chicago, 1970). A political revolution - the violent overthrow of an established polity - is preceded by a crisis. Acute dissatisfaction with the existing regime is so deeply felt by large numbers of people that they risk their lives and property in a desperate attempt to improve their condition. Seeking an analogy, Kuhn postulated a scandal or crisis preceding what he called the "Copernican Revolution":

> The state of Ptolemaic astronomy was a scandal before Copernicus' announcement.
>
> One of the factors that led astronomers to Copernicus...was the recognized crisis that had been responsible for innovation in the first place.
>
> Ptolemaic astronomy had failed to solve its problems; the time had come to give a competitor a chance (pp. 67, 76).

The Alleged Scandal or Crisis before Copernicus. The problems left unsolved by Ptolemaic astronomy were never specified by Kuhn. He referred to "the recognized crisis" without indicating the nature of the crisis or the people who recognized it. What was scandalous about the state of Ptolemaic astronomy before Copernicus?

In the absence of an explicit answer by Kuhn to any of these questions, we must rely on his source, who described as a "scandal" the disagreement between astronomical tables and observations. Was this disagreement particularly scandalous or critical in Copernicus' youth?

The Disagreement between Predictions Based on Astronomical Tables and Actual Observations. In the *Syntaxis* Ptolemy interspersed many tables. Because they were placed near the text discussing their subject matter, they were scattered throughout the thirteen Books. Afterwards he extracted the tables and published them separately as his *Handy Tables*. These were based on a theory that was imperfect. Hence, predictions derived from them tended to diverge from later observations. As time went on, these divergences became more obvious. Hence, post-Ptolemaic astronomers felt the need for new tables. Created by Muslim, Christian, and Jewish astronomers in Asia and Europe, these new tables in their turn diverged from observations. Thus, the authors of the *Alfonsine Tables* said of the *Toledan Tables*:

> Since al-Zarqali's observations two hundred years have passed. In some of the positions which he adopted, there have appeared changes which are obvious and manifest to the senses, so that no excuse can be offered for [retaining] them.[115]

In like manner, as we just saw, Tycho Brahe said about Copernicus that "he found the Alfonsine computations in disagreement with the motions of the heavens."

The Alleged Scandal or Crisis Just Prior to Copernicus Had Been the Norm for Centuries. The scandal or crisis preceding Kuhn's "Copernican Revolution" was no more acute just before Copernicus' announcement than it had been for many centuries theretofore. What Kuhn called a scandal or crisis in Copernicus' youth had been the normal situation for an immensely long time. Kuhn imagined that astronomy had attained a state fundamentally satisfactory to its practitioners - a state that he called a "paradigm"; then the discrepancy between prediction and observation was noticed, creating a scandal or crisis; this was overcome by Kuhn's "Copernican Revolution."

Actually, astronomy had long been, not in a paradigmatic, but an unsettled state. It was not only the discrepancy between prediction and observation that troubled astronomers. They were also baffled by the cosmological or ontological status of the eccentrics and epicycles. Far from being placidly confident about the underpinnings and outcomes of their professional activities, they were at war with one another. Kuhn's concept of paradigm simply does not fit the history of astronomy just before and during Copernicus' time.

How Many Scientific Revolutions? Kuhn developed a generalization from his supposed "Copernican Revolution":

> I shall continue to speak even of discoveries as revolutionary, because it is just the possibility of relating their stucture to that of, say, the Copernican revolution that makes the extended conception seem to me so important (pp. 7-8).
>
> I am repeatedly asked whether such-and-such a development was "normal or revolutionary." And I usually have to answer that I do not know.[116]

In the presence of such uncertainty, we may well ask "How many scientific revolutions?" A noted historian of medicine has said: "The concept of paradigms...does not have any validity in medicine."[117] If we discard paradigms and the purported plethora of revolutions in the history of science, we would do well to recall that the eminent Anglo-Irish physical chemist Robert Boyle (1627-1691) distinguished between two approaches to science: the contemplative and the acquisitive. The replacement of the former by the latter constituted the only Scientific Revolution that has ever occurred. *(See Reading No. 68.)*

Copernicus' Contribution to the Scientific Revolution. Copernicus did not foment a "Copernican Revolution." He did change Ptolemy's stationary earth into a moving planet. He thereby brought science into closer touch with reality. But he never undertook to explain why we on earth do not feel its tremendous motion. Yet by means of planetary parallax he proved that the earth moves. Hence, we know by reasoning what our senses do not report to us. The scope of science was thereby widened: the shortcomings of the senses are remedied by sound reasoning.

In theoretical astronomy, Copernicus discarded Ptolemy's equant. But he kept the Greek's eccentrics and epicycles. These were swept away by the ellipses of that towering Copernican, Johannes Kepler, who began the treatment of heavenly bodies as physical objects. The telescopic discoveries of Galileo and other keen observers were incorporated in the grand consummation achieved by Isaac Newton. When his concepts of absolute space and time came under scrutiny, the relativistic muddle ensued, from which we have not yet emerged.

Notes

108. Bronisław Biliński, *La vita di Copernico di Bernardino Baldi* (Wrocław, 1973), pp. 22-23, lines 103-106, 109-111; Erna Hilfstein, "Bernardino Baldi and His Two Biographies of Copernicus," *Polish Review,* 1979, *24*:75-76.

109. Jacques Quétif and J. Echard, *Scriptores ordinis praedicatorum* (Paris, 1719-1721; reprinted, New York, 1959), II, 126.

110. Rosen, *Three Copernican Treatises,* p. 23.

111. Burmeister, *Rhetikus,* III, 183/17.

112. *Ibid.,* III, 188/27.

113. *Proceedings of the American Philosophical Society,* 1973, *117*:425, 471, 472, 476.

114. *Tychonis Brahe dani opera omnia* (Copenhagen, 1913-1929), I, 149/22-30.

115. Edward Rosen, "The *Alfonsine Tables* and Copernicus," *Manuscripta,* 1976, *20*:164.

116. T. S. Kuhn, "Reflections on my Critics," p. 251, in *Criticism and the Growth of Knowledge,* eds. I. Lakatos and A. Musgrave (Cambridge University Press, 1970).

117. Lester S. King, *The Philosophy of Medicine: the Early Eighteenth Century* (Cambridge, MA, 1978), p. 10.

Part II

Readings

Reading No. 1

Why Is the Moon Eclipsed?

During the Peloponnesian War, Athens sent a major military expedition against Syracuse in Sicily in 415 B.C. The Athenian commander, Nicias, was a timorous man, who was afraid of lunar eclipses. A biography of Nicias was included by Plutarch (c. A.D. 50 - c. A. D. 120) in his ***Parallel Lives.*** *In the following extract* ***(Nicias,*** *¶23) Plutarch contrasts the commander's superstitious fear with Anaxagoras' marvelous discovery.*

The moon was eclipsed at night. Great fear was felt by Nicias and others who trembled at such things through ignorance or superstition. The darkening of the sun around the thirtieth day of the month was already understood by most people as being caused somehow or other by the moon. But what does the moon itself encounter, and how does it lose its light all of a sudden when it is full, and give off all sorts of colors? This was not easy to understand....The first to write an explanation of the moon's brightness and darkness - the clearest and bravest of all the explanations - was Anaxagoras. He was not a man from the past, nor was his teaching accepted....For the physicists and meteorologists, as they were then called, were not tolerated, because they reduced the divine to irrational causes and blind forces and necessary events....Anaxagoras, after being convicted, was barely saved by Pericles [the political leader of Athens].

According to Plutarch's biography of Pericles (¶32), a friend of Nicias, "Diopeithes, proposed a law to prosecute those who did not believe in the gods or promulgated teachings about the heavens, directing suspicion against Pericles through Anaxagoras," who had said the sun was a glowing rock, not a god.

Reading No. 2

The Earth's Shape and Size

Of all the Greek philosophers, the two who influenced later civilization the most were Plato and his pupil Aristotle (384-322 B.C.). Aristotle had an encyclopedic mind, and wrote treatises of lasting value on virtually every subject then known. In his treatise ***On the Heavens*** *(II,14) Aristotle dealt with the shape of the earth and its size.*

All heavy objects fall at like angles [perpendicular to the tangent at the point where the falling object strikes the earth]. The objects, however, are not parallel to one another. This is natural, [their descent being] toward a body spherical in nature....Furthermore, there are also phenomena observed by the senses. For if [the earth were not spherical], the eclipses of the moon would not have such segments [as are observed]. For as a matter of fact in the course of its monthly phases the moon undergoes all the dividing lines (since it becomes straight, convex, and concave). But in the eclipses it always has its boundary line convex. Hence, since it is eclipsed because the earth intervenes, the earth's circumference would be the cause of the [shadow's] shape, the earth being spherical.

Moreover, through observation of the stars it becomes clear that the earth is not only round but also not large in size. For as we move slightly south or north, the circle of the horizon becomes visibly different. Consequently, the stars overhead undergo a big shift, and the same stars do not appear to us as we travel northward or southward. For some stars are seen in Egypt and around Cyprus, whereas they are not visible in the regions to the north. Besides, the stars which are perpetually visible in the northern regions set in those places [farther south]. From these considerations, therefore, it is clear not only that the earth is round in shape, but also that its periphery is not big.

Reading No. 3

Plato's Challenge to the Astronomers

Eudemus, a pupil of Aristotle, wrote a **History of Astronomy,** *of which no copy has come down to us. But before it disappeared, it was used by later writers. These included Sosigenes, an Aristotelian philosopher (2nd century after Christ), who wrote a book on* **The Revolving Spheres.** *Like Eudemus'* **History,** *Sosigenes'* **Revolving Spheres** *perished. But Sosigenes' valuable borrowings from Eudemus were repeated in a work that has survived, the* **Commentary on Aristotle's Treatise On the Heavens,** *written about 540 A.D. by Simplicius. In two passages Simplicius reported Plato's challenge to the astronomers.*

As was related by Eudemus in Book II of his *History of Astronomy,* and Sosigenes took this from Eudemus, Eudoxus of Cnidus is said to be the first of the Greeks to have made use of such hypotheses as those of Plato who, as Sosigenes says, posed the following problem to the students of these subjects: what hypotheses of uniform and regular movements would account for the phenomena involved in the motions of the planets?

It was said previously that Plato, by unequivocally assigning circularity, uniformity and regularity to the heavenly motions, propounded to the astronomers the problem by what hypotheses of uniform, circular, and regular motions it would be possible to explain the planetary phenomena, and that Eudoxus of Cnidus first arrived at the hypotheses of the so-called revolving spheres.[1]

Reading No. 4

Aristotle's Concentric Spheres

Aristotle discussed astronomy not only in his treatise ***On the Heavens*** *but also here and there in his other numerous works. In particular, in his* ***Metaphysics*** *(XII, 8) he explained the concentric spheres devised by Eudoxus, and the improvements introduced by Callippus. Then he proceeded to systematize what these astronomers had done. Instead of treating each planet separately in the manner of Eudoxus and Callippus, Aristotle combined them all in one integrated system. From the highest planet, Saturn, down to the moon, all of Aristotle's heavenly bodies were interconnected.*

If all the spheres in combination are to produce the phenomena, each of the planets must be assigned additional counteracting spheres. Numbering one less [than Callippus' spheres], in all cases they restore the first sphere of the next lower planet to the same position [as it had before being affected by the carrying spheres of the upper planet]. For only in this way can everything together produce the motion of the planets. Of the carrying spheres, there are eight [for Saturn and Jupiter] plus twenty-five [for the sun, moon, and the three other planets]. Of these, only the spheres that carry the lowest body [the moon] do not have to be counteracted. Hence, for the first two bodies [Saturn and Jupiter] the counteracting spheres will be six, and sixteen for the four bodies [Mars, Venus, Mercury, sun] below [Jupiter]. The total of all the carrying spheres and the counteracting spheres that act on them will be fifty-five.

Reading No. 5

The Failure of the Concentric Spheres

During the course of his travels Eudoxus went to Cyzicus in northwestern Asia Minor (now Turkey), and founded a school there. One of his pupils was Polemarchus, later a teacher of Callippus. Because of his association with Eudoxus and Callippus, Polemarchus was familiar with the system of concentric spheres. Even though he was the first to recognize that it could not explain the variation in the brightness and apparent size of the planets, he supported the principle of concentricity. Polemarchus preferred to exalt theory over obser-

vation. Later astronomers, however, realized that when observations conflicted with a theory, it had to be modified or discarded.

The theories of Eudoxus' school do not explain the phenomena. These include not only the phenomena discovered later but also those known previously and recognized by the the Eudoxans themselves....I mean that sometimes the planets appear close by, and at other times remote from us. For this is evident to our eyes in some cases. Thus, Venus and Mars seem many times bigger when they are in the middle of their retrogradations. As a result, on moonless nights Venus makes bodies cast shadows....

The inequality in the relative distances of each [of these bodies] cannot be said to have eluded [the Eudoxans]. For Polemarchus of Cyzicus appears to be aware of it. But he belittles it as imperceptible, because he had a greater liking for the placement of the planets around the center of the universe. (Simplicius, p. 504/17-30, p. 505/19-23)

Reading No. 6

Ptolemy's Enshrinement of the Stationary Earth

Ptolemy, the greatest astronomer of antiquity, lived in the second century after Christ as a Greek-speaking subject of the Roman Empire in the province of Egypt. Of all his numerous works, the most important was the **Mathematical Syntaxis.** *This was called the greatest (***megiste,** *in Greek), by contrast with a group of less important astronomical treatises. Later, when the* **Syntaxis** *was translated by the Muslims, the Arabic definite article* **al** *was prefixed to* **megiste,** *to form "Almagest." This Arabo-Greek hybrid was long used as a misnomer for the* **Syntaxis.** *In Book 1, Chapter 7, it contended that the earth is motionless: it neither revolves in an annual orbit, nor rotates every day about its axis.*

That the Earth Performs No Progressive Motion

In the same way it will be proved by what precedes that the earth cannot make a contrary motion to the aforementioned lateral sides, or ever be displaced at all from its position in the center [of the universe]...The earth occupies the central position in the cosmos, and all heavy objects move toward it....If it had any one movement in common with the other heavy bodies, it would outstrip them all in its descent because its size is so much bigger. It would leave living creatures behind, and partly dense bodies floating on the air. For its part, it would swiftly drop out of the heavens altogether....

Yet without having any objection to these views, some people agree on what they regard as more plausible. They think nothing would contradict them if, for example, they supposed the heaven motionless while the earth whirled around its own axis from west to east every day....As regards the heavenly phenomena, perhaps nothing would prevent the situation from being in agreement with the simpler arrangement. But they failed to notice that as far as what happens around us and in the air is concerned, such a view would appear quite ridiculous.

For let us grant them what is unnatural: the lightest and least dense bodies [the stars] do not move at all...while the densest and heaviest bodies execute their own swift and uniform motion....The geokineticists would then admit that of all the motions in the earthly region the swiftest is the rotation of the earth. In a short time it performs so vast a rotation. As a result, everything not attached to it would appear to be always moving in the direction opposite the earth's. Not a cloud nor any other thing flying or thrown upward would ever be seen floating eastward. For the earth would always outstrip them all as it overtook them in its eastward motion. Consequently all other bodies would seem to be traveling westward as they were left behind.

On the other hand, the geokineticists might say that the air is carried around with the earth in the same direction and at the same speed. Nevertheless, whatever bodies are caught up in the air would always be seen falling behind the speed of both the earth and the air. Or if these bodies were carried around as though united with the air, they would no longer be seen either moving ahead or falling behind. On the contrary, they would always maintain the same position. Neither in the things flying or thrown would there be any dislodgment or displacement. [Yet] we do see all these things happening with such clarity that no part at all of their slowness or swiftness is attributable to the earth's failure to stand still.

Reading No. 7

Copernicus' Intellectual Indebtedness to the University of Cracow

Albert Caprinus received the bachelor of arts degree from the University of Cracow in the spring of 1541. On 27 September 1542 he dedicated his **Astrological Forecast (Iudicium astrologicum)** *for the year 1543. As a passionate partisan of his alma mater, in his dedication Caprinus proudly proclaimed (and somewhat exaggerated) Copernicus' intellectual indebtedness to the University of Cracow.*

At this university mathematics was imbibed by many men whose teaching of that subject in Germany is highly praised and beneficial to students. By way of honorable mention among them[2] I name Nicholas Copernicus, the canon of Varmia. He once enjoyed the hospitality of this city. His wonderful writings in the field of mathematics, as well as the additional materials he has undertaken to publish, he first acquired at this university of ours as his source. He does not deny this, in agreement with Pliny's judgment that "to identify those from whom we have benefited is an act of courtesy and thoroughly honest modesty." Moreover, whatever this benefit, he admits that he received all of it from our university.

Reading No. 8

Copernicus' Acceptance of Eccentrics and Epicycles

Copernicus was convinced that the sun and the stars do not move. Yet they appear to move, since we look at them from the earth which is in motion. We do not feel the earth's motion because we share in it. The sun and stars do not actually rise and set. The earth's daily rotation, which we do not feel, makes it appear that the sun and the stars rise and set. "Let nobody be surprised," said Copernicus in the introduction to Book II of the **Revolutions**, *"if I still refer simply to the rising and setting of the sun and stars, and similar phenomena. On the contrary, it will be recognized that I use the customary terminology, which can be accepted by everybody." Thus Copernicus often spoke of the sun's apparent motion when he was referring to the earth's annual revolution.*

The sun's [apparent motion], however, is demonstrably nonuniform, because the motion of the earth's center in its annual revolution does not occur precisely around the center of the sun. This can of course be explained in two ways, either by an eccentric circle, that is, a circle whose center is not identical with the sun's center, or by an epicycle on a concentric circle [that is, a circle whose center is identical with the sun's center, and which functions as the epicycle's deferent]....Whatever is done by an epicycle, however, can be accomplished in the same way by an eccentric. This is described equal to the concentric and in the same plane by the planet as it travels on the epicycle, the distance from the eccentric's center to the concentric's center being the length of the epicycle's radius....The same apparent nonuniformity always occurs either through an epicycle on a concentric or through an eccentric equal to the concentric. There is no difference between them, provided that the distance between their centers is equal to the epicycle's radius. Hence it is not easy to decide which of them exists in the heavens.[3]

Reading No. 9

Copernicus' Theory of Motion

Aristotle assigned upward motion to air and fire, the two lightest of his four elements. To the two heaviest elements, water and earth, he assigned downward motion. These four elements were found in the region below the heavens, where only circular motion occurred. All these heavenly motions were centered on the earth, which lay motionless in the middle of Aristotle's universe. Copernicus proclaimed, however, that the earth is a planet. Like the other planets, it revolves around the center of the universe. As it performs this circular revolution, heavy bodies still fall down, and light bodies rise.

The motion of falling and rising bodies in the framework of the universe is twofold, being in every case a compound of straight and circular. For, things that sink of their own weight, being predominantly earthy, undoubtedly retain the same nature as the whole of which they are parts. Nor is the explanation different in the case of those things, which, being fiery, are driven forcibly upward. For also fire here on the earth feeds mainly on earthy matter, and flame is defined as nothing but blazing smoke. Now it is a property of fire to expand what it enters. It does this with such great force that it cannot be prevented in any way by any device from bursting through restraints and completing its work. But the motion of expansion is directed from the center to the circumference. Therefore, if any part of the earth is set afire, it is carried from the middle upwards. Hence the statement that the motion of a simple body is simple holds true in particular for circular motion, as long as the simple body abides in its natural place and with its whole. For when it is in place, it has none but circular motion, which remains wholly within itself like a body at rest. Rectilinear motion, however, affects things which leave their natural place or are thrust out of it or quit it in any manner whatsoever. Yet nothing is so incompatible with the orderly arrangement of the universe and the design of the totality as something out of place. Therefore rectilinear motion occurs only to things that are not in proper condition and are not in complete accord with their nature, when they are separated from their whole and forsake its unity.

Furthermore, bodies that are carried upward and downward, even when deprived of circular motion, do not execute a simple, constant, and uniform motion. For they cannot be governed by their lightness or by the impetus of their weight. Whatever falls moves slowly at first, but increases its speed as it drops. On the other hand, we see this earthly fire (for we behold no other), after it has been lifted up high, slacken all at once, thereby revealing the reason to be the violence applied to the earthy matter. Circular motion, however, always rolls along uniformly, since it has an unfailing cause. But rectilinear motion has a cause that quickly stops functioning. For when rectilinear motion brings bodies to their own place, they cease to be heavy or light, and their motion ends. Hence, since circular motion belongs to wholes, but parts have rectilinear motion in addition, we can say that "circular" subsists with "rectilinear" as "being alive" with "being sick." Surely Aristotle's division of simple motion into three types, away from the middle, toward the middle, and around the middle, will be construed merely as a logical exercise. In like manner we distinguish line, point, and surface, even though one cannot exist without another, and none of them without body.[4]

Reading No. 10

The Brothers Copernicus Borrow Money from a Bank

Lucas Watzenrode, bishop of Varmia, was engaged in a bitter conflict with his neighbors, the Knights of the Teutonic Order. This struggle's legal issues were debated in Rome. For this purpose, Watzenrode sent his secretary, George Pranghe, to Italy. When Pranghe reached Bologna, he was sought out by the bishop's nephews, Nicholas Copernicus and his older brother Andrew. They were both students of law at the University of Bologna, and had run out of money. Pranghe himself could not help them, because his cash was exhausted. But when he arrived in Rome, he explained the brothers' predicament to Bernard Scultetus. As dean of the Varmia Chapter, Scultetus was stationed at the papacy to look after the interests of the Varmia Chapter, to which the brothers Copernicus belonged as two of its sixteen canons. Scultetus guaranteed to repay a bank 100 ducats borrowed by the brothers Copernicus on or about 21 September 1499. A month later, on 21 October 1499, Scultetus wrote to Watzenrode, asking him to deposit the 100 ducats, plus accumulated interest, either in Poznań or Wrocław, for transmission to Rome. The part of Scultetus' letter pertaining to the brothers Copernicus is translated below.

> Your Reverence's nephews, who live in Bologna, were short of money these past days, in the manner of students. They betook themselves to George [Pranghe, secretary of Copernicus' uncle, Lucas Watzenrode, the bishop of Varmia], as truly destitute to someone destitute, asking him what advice he had. Andrew proposed to seek employment in Rome, to alleviate his poverty. They finally received a hundred ducats from a bank, charging interest. I guaranteed to repay the loan in four months, of which one has already passed. Hence, to avoid further loss to your nephews, and impairment of credit-

worthiness to me as guarantor, I humbly ask your Reverence not to disdain to deposit the aforesaid sum as quickly as possible in Poznań or Wrocław for transmission to Rome. In this matter your Reverence will perform a favor advantageous to your nephews and most pleasing to me, a favor to be returned to your Reverence in equal measure without fail.

Rome, 21 October 1499

Your Most Reverend Excellency's
humble servant
Bernard Scultetus
Dean of Varmia[5]

Reading No. 11

The Varmia Statute Requiring Three Years of University Study

The Varmia Chapter, to which Nicholas and Andrew Copernicus belonged as canons, was governed by statutes, which had been revised about a decade before the brothers Copernicus were inducted. Section 51 of those revised statutes dealt with the education of the canons.

The [Varmia] church needs to be staffed with educated people, capable of producing seasonable fruit at the right time. We have therefore decreed that every newly admitted canon, unless he has a master's or bachelor's degree in Biblical studies or canon or civil law, or holds a doctorate or license in medicine or the healing art, after his first year's residence, if it is agreeable and convenient for the Chapter, is required to study at least three years in some authorized university in one of the aforementioned faculties. He shall devote his energies solely to his studies so as to persevere continuously and without interruption in them throughout the aforementioned period of three years.

He is not to presume to absent himself from the institution, unless he decides that he should transfer to another recognized university on account of plague, illness, famine or war. If, however, he leaves the university on any other ground, provided it is reasonable, during the time of his absence the Chapter will at the very least treat him as a student. But if he decides to absent himself for a frivolous reason (whether this is so, will be settled by the judgment of the Chapter), during the time of such absence he shall simply be deemed absent. Nevertheless, he shall be required to begin the aforesaid three-year term again, and continue his studies, as previously indicated, to the very end, just as if he had done nothing in them previously.

Therefore, with regard to each and all of the aforementioned cases, before a canon on his return is admitted to receiving [his share of] the distribution [of the canons' income], he shall be required to furnish full faith by means of an open letter, certified by the seal of the rector of the university in which he studied, and by the swearing of his own oath.

Moreover, if he conducted himself profitably in the university, and requests that he be permitted to study longer, that permission will not be withheld from him.[6]

Reading No. 12

The Varmia Chapter Grants the Brothers Copernicus Permission to Study

The brothers Copernicus, Nicholas and Andrew, left Italy and returned to Varmia to obtain the Chapter's approval of their plans to study. They both appeared before the Chapter on 27 July 1501. Yet there is a striking difference between the two brothers. Nicholas stated that he had already spent three years in study with the Chapter's permission. Andrew, on the other hand, requested approval to begin his studies. Yet he had been enrolled in the University of Bologna as a student of law. He had evidently done so without procuring the Chapter's consent in advance. Nicholas based his request on his intention to study medicine, and on his promise to treat the canons professionally. This particularly pleased the Chapter, none of whose members had medical training. On the other hand, it judged that Andrew "also seemed qualified to engage in studies."

Nearly two months later the Chapter met to consider a third canon's request for a leave of absence for purposes of study:

In the same year as above, on 23 September Henry Nidderhoff, canon of Varmia, requested permission to study in the Roman Curia for two years. The members of the Chapter read him Section 50 [should be 51] of the statutes as contrary and opposite to his request. Henry pressed his petition more insistently, arguing that the permission previously granted to Andrew Copernicus should not be denied to him. After long consideration the Chapter finally agreed to his request.[7]

Since the Chapter permitted Andrew Copernicus to attend a university, why should it withhold such permission from any other canon? No argument of this kind was directed against Nicholas Copernicus.

In the year 1501,[8] on martyr Pantaleon's day [27 July], before the Chapter there appeared Canons Nicholas and Andrew Copernicus, who are brothers. Nicholas had already spent three years in a university with the Chapter's permission. He requested an extension of his period of study, namely, for two years. The other brother,

Andrew, asked for approval to begin his period of study and continue it in keeping with the spirit of the statutes, with each brother being given what is usually granted to students.

After thorough consideration the Chapter acquiesced in the wishes of both brothers, particularly because Nicholas promised to study medicine with the intention of advising our most reverend bishop in the future, and also the members of the Chapter, as a healing physician. Andrew also seemed qualified to engage in studies.[9]

Reading No. 13

Copernicus' Scholastry in Wrocław

The budget of the Church of the Holy Cross in Wrocław (formerly Breslau) provided for an official, called the scholaster, who was expected to supervise the instruction given there. A vacancy occurred in this scholastry while Copernicus' uncle was the bishop of Varmia, who possessed the right to nominate a candidate to fill this vacancy. At that time Copernicus was studying medicine at the University of Padua. When he was notified that he had been appointed scholaster in Wrocław, he could not leave Padua to take possession in person of his new position. Hence he resorted to the legal device of a proxy, naming two Wrocław canons to act on his behalf. Accompanied by two witnesses, he hurried on 10 January 1503 to the office of a public notary in Padua. There he jotted down a draft of the proxy, to make things easier for the notary. But he was in such a rush that he committed no less than five errors in the nine lines of his draft. This document, which has been preserved in the notary's files in the Paduan Archive, is the oldest surviving example of Copernicus' handwriting.

> I, Nicholas Copernik, canon of Varmia and scholaster of the Church of the Holy Cross in Wrocław, revoking [whatever proxies I have constituted heretofore in any way,] designate as proxies the honorable man Apicius Colo, chancellor and canon of the cathedral of Wrocław, and Michael Jode, canon of the same cathedral of Wrocław, for the purpose of taking possession of the said scholastry, recently conferred on me, and whatever other [collations and provisions have been made or are to be made for the same designator, from whatever ecclesiastical benefices, wherever constituted, by apostolic or episcopal or any other authority, and for initiating and sending appropriate letters, and presenting them to executives, and seeking and obtaining appropriate procedures for decisions about those letters, and also presenting, publishing, publicizing, and exhibiting the letters and procedures to all and sundry as needed, and ad-

vising and requesting them and each of them to heed and obey the letters and procedures under the penalties and censures contained therein, and whatever benefits under whatever grants made and to be made to him etc.].

Concerning all and each of these matters, the same Nicholas, acting for himself, asked me, the undersigned notary public, to make and draw up one or more public instrument or instruments.

These [instruments] were drawn up in Padua, in the chancellery of the bishop of Padua, in the year, indiction, month, day, and pontificate, as above, while there were present in the same place the venerable Leonard Rodinger of the diocese of Passau, and Nicholas Monsterberg of the diocese of Włocławek, as accepted witnesses to the foregoing, especially called and summoned.[10]

Reading No. 14

Copernicus' Doctoral Diploma

The statutes of the Varmia Chapter, to which Canon Copernicus belonged, required every canon who had been granted a leave of absence for the purposes of study to return with a degree from a recognized university. Copernicus had enrolled in two of the most famous Italian universities, Bologna and Padua. But when his leave of absence was about to expire, and the time had come for him to go back to Frombork, he preferred to take his doctoral degree neither at Bologna nor at Padua. At those two prestigious centers of learning the costs for a doctoral candidate were quite steep. Copernicus chose instead the less renowned but also less expensive University of Ferrara. There on 31 May 1503 he received a doctoral diploma in canon law.

1503, on the last day of the month of May, in Ferrara, in the bishop's palace, beneath the balcony of the garden, in the presence of witnesses summoned and invited, the honorable Giovanni Andrea Lazari of Palermo in Sicily, the distinguished rector of the gracious school of law in Ferrara; Bartolomeo Silvestri, citizen of Ferrara and notary; Lodovico, son of the late Baldassarre, Regio, citizen of Ferrara and beadle of the law school of the city of Ferrara; and others;

In the name of Christ, the venerable and very learned gentleman, Nicholas Copernich of Prussia, canon of Varmia and scholaster of the Church of the Holy Cross in Wrocław, who studied at Bologna and Padua, was approved in canon law, with absolutely nobody in disagreement, and given the doctoral degree by the aforesaid Giorgio [Prisciani,] the aforementioned suffragan [bishop of Ferrara].

The sponsors were Filippo Bardella and Antonio Leuti, citizens of Ferrara etc., who gave the insignia to the candidate.[11]

Reading No. 15

Copernicus' Medical Bonus

Copernicus was permitted by the Varmia Chapter to leave Frombork and join the staff of his uncle, the bishop of Varmia, in Lidzbark. But when the bishop fell ill, at a meeting on 7 January 1507 the Chapter resolved to give Copernicus a bonus of fifteen marks every year, over and above his regular income as a canon. The Chapter took this action especially because Copernicus was supervising his uncle's recovery. Copernicus practiced medicine even though he had not studied long enough at the University of Padua to earn an M.D. The medical profession was not as overcrowded then as it has since become.

In the year 1507, on 7 January, Nicholas Kopernig, our fellow canon, having been released to serve our most reverend ruler [the bishop of Varmia], received by the exceptional good-will of the Chapter, in addition to the income from his office, 15 marks of the good coinage, to be paid every year until he withdraws from the staff of the bishop. This bonus is granted to him with pleasure, especially since he excels in the art of medicine, and with his care and remedies sagely supervises the recovery of His Reverence.[12]

Reading No. 16

Copernicus' Service to the Varmia Chapter as Inspector

In 1510 Copernicus decided to leave his uncle's episcopal palace in Lidzbark and rejoin the Varmia Chapter in Frombork. Together with a fellow canon, who not long thereafter succeeded his uncle as bishop of Varmia, Copernicus was named inspector of the Chapter's holdings. According to Section 38 of the Chapter's statutes,

Every year at the Chapter's general meeting, which is usually

> *held around All Saints' Day [1 November], two of the canons shall be designated to collect the rents in Mehlsack and Olsztyn. Together with the Administrator of the Chapter, they shall properly and carefully inquire of the judges and the two heads of the individual villages, at the time when they are paying the rents, or of any other inhabitants of the aforesaid areas, about all the deficiencies in individuals as well as in the condition of the area. The inspectors shall undertake to make suitable changes as far as they can, but more serious matters shall be referred to the Chapter for regulation. But if perhaps the two aforementioned canons, or either of them, are or is lawfully prevented from being able to perform this obligation at the aforesaid times, the Chapter shall have the right to substitute another or others in [his or] their place.*[13]

Copernicus was not prevented in any way from performing his duty as inspector.

In the year 1511, by order of the Venerable Chapter, we, Fabian of Lossainen and Nicholas Koppernig, appointed inspectors by the Venerable Chapter, in Olsztyn on the day of the Lord's Circumcision [1 January], received the remaining money for the vicariates of the venerable Zacharias [Tapiau[14]] on deposit in the fortress, namely, 238 3/4 marks.

By order of the Chapter, we handed this money to the Venerable B[altasar] Stockfisch on our return to the cathedral[15] [in Frombork].

Reading No. 17

Copernicus as the Target of Evil Gossip

A woman who worked as Copernicus' housekeeper left to get married. Her husband turned out to be impotent. She separated from him, and found employment with a matron in Elbląg. Together they visited the Koenigsberg fair. On the way back, they had to spend the night in Frombork. In recognition of his former housekeeper's loyal services, Copernicus permitted the two women to spend the night in his home. The episode was reported to the bishop, who admonished Copernicus. On 27 July 1531 Copernicus replied.

The original has not been found, but a late copy is preserved. This was transcribed by an editor of Copernicus' works. He refrained from publishing it, however, because it reveals Copernicus' recognition of the "bad opinion" arising from his conduct. He promised the bishop that in the future nobody would have any valid reason to circulate evil gossip about him.

My lord, Most Reverend Father
in Christ, my noble lord:

With due expression of respect and deference, I have received your Most Reverend Lordship's letter. Again you have deigned to write to me with your own hand, conveying an admonition at the outset. In this regard I most humbly ask your Most Reverend Lordship not to overlook the fact that the woman about whom your Most Reverend Lordship writes to me was given in marriage through no plan or action of mine. But this is what happened. Considering that she had once been my faithful servant, with all my energy and zeal I endeavored to persuade them to remain with each other as respectable spouses. I would venture to call on God as my witness in this matter, and they would both admit it if they were interrogated. But she complained that her husband was impotent, a condition which he acknowledged in court as well as outside. Hence my efforts were in vain. For they argued the case before his Lordship the Dean [of the Chapter], your Very Reverend Lordship's nephew, of blessed memory, and then before the Venerable Lord Custodian [of the Chapter]. Hence I cannot say whether their separation came about through him or her or both by mutual consent.

However, with reference to the [present] matter, I will admit to your Lordship that when she was recently passing through here from the Koenigsberg fair with the woman from Elbląg who employs her, she remained in my house until the next day. But since I realize the bad opinion of me arising therefrom, I shall so order my affairs that nobody will have any proper pretext to suspect evil of me hereafter, especially on account of your Most Reverend Lordship's admonition and exhortation. I want to obey you gladly in all matters, and I should obey you, out of a desire that my services may always be acceptable.

Frombork, 27 July 1531
Your Most Reverend Lordship's

most devoted
Nicholas Copernicus

To his lordship, Most Reverend Father
in Christ, Maurice [Ferber], by the
grace of God bishop of Varmia,
my gracious and most honorable lord[16]

Reading No. 18

Copernicus' Letter about his Housekeeper

When John Dantiscus was appointed bishop of Varmia, he toured the diocese in the company of two canons, Copernicus and Felix Reich. Speaking privately to Copernicus about the general problem of the canons' employment of housekeepers, Dantiscus advised him to make a change. Copernicus had trouble finding a qualified female relative at once. He hoped he could do so by Easter, 6 April 1539. But Dantiscus demanded earlier compliance. On 2 December 1538 Copernicus promised to settle the matter by Christmas of that year.

My lord, Most Reverend Father in Christ,
most gracious lord, to be heeded by me
in everything:

I acknowledge your Most Reverend Lordship's quite fatherly, and more than fatherly admonition, which I have felt even in my innermost being. I have not in the least forgotten the earlier one, which your Most Reverend Lordship delivered in person and in general. Although I wanted to do what you advised, nevertheless it was not easy to find a proper female relative forthwith, and therefore I intended to terminate this matter by the time of the Easter holidays. Now, however, lest your Most Reverend Lordship suppose that I am looking for an excuse to procrastinate, I have shortened the period to a month, that is, to the Christmas holidays, since it could not be shorter, as your Most Reverend Lordship may realize. For as far as I can, I want to avoid offending all good people, and still less your Most Reverend Lordship. To you, who have deserved my reverence, respect, and affection in the highest degree, I devote myself with all my faculties.

Gynopolis, 2 December 1538
Your Most Reverend Lordship's

most obedient
Nicholas Copernicus

To his lordship, Most Reverend Father
in Christ, Johannes [Dantiscus],
by the grace of God bishop of Varmia,
his most gracious lordship[17]

Reading No. 19

Felix Reich Balks at Reading Dantiscus' Letter Aloud to Copernicus

When Bishop Dantiscus of Varmia decided to reprimand Copernicus in the matter of his housekeeper, the bishop entered into a secret agreement with Felix Reich, a fellow canon of Copernicus. Dantiscus was to send a letter to Reich, who would read it aloud to Copernicus in order to intensify the pressure on the astronomer to dismiss his housekeeper right away. But in his anger the bishop imprudently inserted "certain little words," which deterred Reich, a professional notary, from carrying out his part of the agreement.

My lord, Most Reverend Father
in Christ, most gracious lord:

After the most dutiful expression of my devotion, I give voice to undying thanks to your Most Reverend Lordship for taking so fatherly an interest in my illness. May the Lord, who rewards all good will, repay you for that kindness and benevolent disposition.

As far as the venerable Nicholas Copernicus is concerned, I approve of your Most Reverend Lordship's pious initiative and fatherly admonition. I hope that he will take it to heart, so as not to need my admonition. He will be overcome with shame, I am afraid, if he learns that I am privy to this matter. Had I not been prevented by the insertion of certain little words, I would perhaps have read to him your Most Reverend Lordship's letter insofar as it touches on that business, for this was agreed between us. Hereafter (as your Reverence will recognize from the enclosed copy) it will be clear from his reply to what direction the affair will turn....

Frombork, in great excitement,
2 December 1538[18]

Reading No. 20

Reich's Writ against Copernicus

Felix Reich's health was so bad that his life was saved by Copernicus, who as the attending physician stopped the loss of blood. Nevertheless, after his recovery Reich, a professional notary, drafted two writs, one directed against Copernicus' housekeeper, and the other against Copernicus himself. Reich sent these writs to Dantiscus. On the basis of the earlier writ, Dantiscus could have the local priest issue a warning to the housekeeper. She had a husband, unlike the two other women involved in this situation. In accordance with Reich's later writ, the bishop could issue an order at the earliest possible moment to be sent sealed to Copernicus, and to each of the two other canons under attack.

With regard to another matter, I observe that your Most Reverend Lordship is anxious to remove the serious scandal in the church. Hence I have no doubt that a prosecution too will be forthcoming, and that other upright canons are waiting for this [move] with keen desire. I therefore now, without being asked and of my own volition, step forward to impart gladly whatever I can muster, as long as I can, by way of advice or service for so pious a proceeding.

It therefore seems to me that your Most Reverend Lordship should send in a sealed letter at the earliest possible opportunity individually to each of these three of our brothers [one of whom was Copernicus] an order in accordance with the writ dictated by me and written down - since I myself can barely write - by the faithful hand of Fabian [Emerich (1477-1559), the Chapter's notary], acting as secretary, who may be safely entrusted with everything. And in addition the females should likewise be warned by the local priest under your authority in accordance with the writ which I sent previously. Care should also be taken to omit from the letters to the other two, who do not have legal husbands, what is in that earlier letter concerning Nicholas [Copernicus'] cook, who does have a legal husband. The impending commencement of the proceeding against the women too will strike terror to no small degree. Yet an appropriate limit must be granted to them so that they may in all likelihood be able to secure other homes for themselves within the boundaries of the warning. Have the venerable Achatius [von der Trenck], the Administrator in Olsztyn, inform your Most Reverend Lordship about what is generally said in this regard. Yet he is privy to none of the matters about which I am writing.

Whatever the situation may be, may your Most Reverend Lordship act firmly. God Almighty will strengthen your arm so that you may conduct to a happy ending what you initiated out of zeal. As much as we can, all of us will help make a success of this affair. However, your Most Reverend Lordship must take care nevertheless in commencing the proceeding with the force of law not to introduce in your future letters anything contrary to formal and customary legal style, as it is called. For it often happens that even the tiniest clause may spoil an entire case, so that it is declared null and void if it comes before a higher judge.

I commend myself to the grace and favor of your Most Reverend Lordship. May you be safe and sound for a long time!
Frombork, 11 January 1539
Your Most Reverend Lordship's

F[elix] R[eich]

To my lord, Most Reverend Father in Christ,
Johannes [Dantiscus], by the grace of God
bishop of Varmia, my most worshipful
and gracious lord[19]

Reading No. 21

Reich Edits Dantiscus' Documents

The legal documents drafted by Dantiscus and his staff against three canons and their housekeepers were sent to Reich for editorial scrutiny. He found them defective, suggested corrections, and returned them to Dantiscus on 23 January 1539. He also indicated how they should be sent, after they had been put in their final form. The letters to the three women should be tied together in a single bundle, addressed to the priest. The courier should be told to deliver this bundle first, before any delivery to the canons. For if the canons intercepted the letters intended for the women, those letters would surely not achieve what Dantiscus desired.

In writing the address on the outside of this letter, Reich mistakenly called Dantiscus "bishop of Chełmno," the office vacated by Dantiscus the year before, when he moved up to the more lucrative bishopric of Varmia. Reich wrote this letter at night. Was he afraid that he would be unable to function on the following day? His physicians have made their nearly final decision. He was down to his last flagon of wine. He apologized to Dantiscus for not being "better organized and clearer and more comprehensive."

Most Reverend Father in Christ,
Most gracious lord:

I am sending back all the letters because in one a serious scribal error must be corrected, and that cannot be done here. For the scribe wrote "Henry" instead of "Alexander" [Scultetus, one of the two other Varmia canons under attack, in addition to Copernicus]. Moreover, in my previous letter I warned about the banishment of ten miles and outside the diocese, since your Most Reverend Lordship does not have the power to banish anyone beyond your own diocese, which in certain places (as here [Frombork]) does not extend farther than one mile. Consequently it would have been necessary to delete this reference to ten miles as the distance to which the women are relegated. It is my advice that this too should be done now. Finally, "innocent" is written elsewhere instead of "behaving innocently." And in case there are any other [defects], I am for that reason sending all the documents back as a group, so that also as a group they may hereafter be put in final form one by one.

It is also necessary that the letters to the canons should be tied up separately, and that in like manner the letters to the cooks should be tied up separately, sealed in one envelope, and addressed to the priest. For if this were not done, a great mishap could occur. For if open letters to the cooks fell into the hands of those three canons, I have no doubt that, being intercepted, the letters could not accomplish their purpose. Your Most Reverend Lordship will therefore instruct your courier on his return [from Lidzbark to Frombork] to deliver to the priest the documents pertaining to the women first and ahead of everything [else], and afterwards the documents concerning the canons to any canon. The latter will undoubtedly give each one his own and, if need be, accost each one [uncertain reading] most circumspectly. Otherwise he will heap no small suspicion on me.

I suppose your Most Reverend Lordship has a sound reason for writing very briefly to Nicholas [Copernicus], and it does not matter a great deal. Undoubtedly they all [Copernicus and the two other canons under attack] coordinate everything with one another.

As regards the beverage I shall request from the generosity of your Most Reverend Lordship, I have previously designated as my agent the venerable Administrator in Olztyn. In nearly their final decision my physicians now very discreetly allow wine in unlimited quantities to strength my heart, a pure, clarified, and gentle wine, however, not harsh or too sweet or too strong [illegible reading], but a mild Hungarian wine, muscatel, malmsey or suchlike. One flagon of a wine of this category will be enough for me, I believe. The Piotrków beer which your Most Reverend Lordship gave me, I now for the first time feel is very wholesome. At present, therefore, I drink it every day. No small quantity is still left, however, so that there is absolutely no need for concern on the part of your Most Reverend Lordship, to whom I wish to be of the greatest service. Better organized and clearer and more comprehensive, I cannot be. May Christ keep your Most Reverend Lordship safe.
Frombork, 23 January 1539 at night
Your Most Reverend Lordship's

F[elix] R[eich]

To my lord, Most Reverend Father in Christ,
Johannes [Dantiscus], by the grace of God
bishop of Chełmno, my lord and
most gracious superior[20]

Reading No. 22

Reich Delays the Delivery of Dantiscus' Letter to the Varmia Chapter

Dantiscus wrote a letter addressed to the Varmia Chapter, which he sent to Reich for delivery to the Chapter. Since the letter was not addressed to Reich, he would not open it. Hence, he could not be sure whether or not it concerned the bishop's proceedings against Copernicus and two other canons. If such were its contents, an uproar might result, with Reich being exposed as a collaborator with Dantiscus in his attack upon three fellow canons of Reich. He wanted to keep his dealings with Dantiscus a secret. Hence, he seized upon the small number of canons who happened to be in residence at the moment as a pretext for sending the letter back to Dantiscus. It could be redirected to the Chapter without any great loss of time.

My lord, Most Reverend Father
in Christ, most gracious lord:

Today I received the wine and the Masovian beer, which however smells strongly of lavender, and that is incompatible with my illness. Nevertheless, I am immensely grateful for everything. But I beg your most Reverend Lordship not to be concerned about additional beer, since the previous [supply] has not yet been exhausted. As soon as it gives out, I shall not refrain from calling once more on your Most Reverend Lordship's generosity.

Together with the wine from Olsztyn, through the effort of the venerable Administrator in Olsztyn your letter to the Chapter was also delivered to me. I am afraid, however, that it may contain something about the proceedings against the canons' cooks and against the canons themselves. I do not dare deliver the letter to the Chapter lest its members cause a disturbance in this affair. It would be expedient to have the matter settled as soon as possible, and have the letter sent back. Whatever your most Reverend Lordship may then order regarding the letter to the Chapter, I shall see to it that the order is carried out. Meanwhile I beg you not to be angry because without any instructions I withheld the letter in accordance with my own judgment. The loss of time will be slight. For, momentous matters cannot be handled now by only three members of the Chapter [the others being temporarily absent].

I commend myself to the customary grace and favor of your Most Reverend Lordship forever.
Frombork, 27 January 1539
Your Most Reverend Lordship's

Felix Reich
with sick and trembling hand

To my lord, Most Reverend Father in Christ,
Johannes [Dantiscus], bishop of Varmia,
my lord and most gracious superior[21]

Reading No. 23

Provost Płotowski Reports to Bishop Dantiscus about the Housekeepers of the Three Canons under Attack

Paul Płotowski, the first Pole admitted to the Varmia Chapter, was appointed Provost of the Chapter by the king of Poland. To ingratiate himself with Bishop Dantiscus, who had previously served the Polish king as a diplomat, Płotowski sent a report about the women connected with the three canons under attack: Nicholas Copernicus, Alexander Scultetus, and Leonard Nidderhoff.

Most Reverend Father in Christ...I have not written to your Most Paternal Reverence to the same extent about the women in Frombork. The one connected with Alexander [Scultetus] hid herself in the house for several days out of fear of the second decision. I told her to leave after obtaining possession of one child, which was not what she sought [uncertain reading]. Alexander [Scultetus] came back from Lubawa [Löbau, in German] with a happy countenance. I do not know what he reported. He is staying with [Leonard] Nidderhoff and his housekeeper in their lodgings, while she acts like a matron brewing beer, oblivious of all her troubles. The one connected with Doctor Nicholas [Copernicus] sent her things ahead to Gdańsk [Danzig, in German]. Yet she is still staying by herself in Frombork.

It would be a good idea for your Paternal Reverence to entrust the office supervising the vicars and the position in charge of the liquors to the venerable Custodian [John] Timmermann...for I do not know nor do I hear that anyone is better qualified than he is. Of course, your Reverence has men with doctoral degrees in your church. But whither their studies lead them, your Paternal Reverence knows better than I can write....

Frombork, 23 March 1539[22]

Reading No. 24

Tiedemann Giese Denies Having Encouraged Scultetus

On 23 March 1539 Provost Płotowski wrote to Bishop Dantiscus that Alexander Scultetus had returned to Frombork from Lubawa "with a happy countenance." Lubawa was the see of the diocese of Chełmno (Kulm, in German), where Tiedemann Giese (1480-1550) was the presiding bishop. After receiving Płotowski's report about Scultetus, Dantiscus wrote very promptly to Giese, asking what made Scultetus so happy in Lubawa. Giese replied equally promptly, denouncing Płotowski's craftiness.

I laughed when I read what your Reverence writes about Alexander [Scultetus] after he returned from [visiting] me. I marveled at the craftiness of the informer. For a long time he has been angry with me for refusing to take his side when your Reverence was aroused against him. Now that he has been taken back in your good graces, he thinks he can get revenge if he can stir up some quarrel between us. If I wished, I could pay him back most generously. I talked plainly to Alexander [Scultetus]...he left me in despair....Hence I do not know what joy he could have brought home.

Chełmno, 4 April 1539[23]

Reading No. 25

Giese as Copernicus' Patient

While visiting an area in his diocese, Bishop Giese of Chełmno fell ill. Returning to his episcopal residence in Lubawa, he was treated by two outside physicians. Copernicus also was summoned from Frombork. He arrived in Lubawa on 27 April 1539, as was mentioned by Giese's chaplain reporting to Dantiscus about Giese's illness.

...The bishop received medications from Doctor Jerome of Toruń which had been left for him, and also other drugs, unfamiliar to me, from Doctor Ambrose of Gdańsk....The doctors promise an improvement in the future, not only the aforementioned physician from Gdańsk but also Nicholas Copernicus, canon of Varmia, who arrived here today [27 April 1539].[24]

Reading No. 26

Dantiscus Asks Giese to Persuade Copernicus to Avoid Anna Schilling and Alexander Scultetus

Having been informed of an impending visit by Copernicus to Giese, Dantiscus hurriedly wrote a letter to the bishop of Chełmno on 5 July 1539. This letter began with Dantiscus' claim to be an old friend and great admirer of Copernicus. He was aware, however, that in actuality Giese was the astronomer's closest friend. Hence, Dantiscus urged Giese to give Copernicus two pieces of advice. In the first place, the astronomer, then over sixty-six years old, "is said to let his mistress in frequently in secret assignations." Secondly, Copernicus allows himself to be led astray by his fellow canon Alexander Scultetus, who is suspected of sympathizing with the Protestant Reformation. Dantiscus wanted Giese to convince Copernicus to stop seeing Anna Schilling and Scultetus. But Dantiscus wished this advice to sound to Copernicus as though it originated with Giese, who was not responding to any suggestion from Dantiscus.

I have heard that your Reverence will receive the distinguished and very learned man, Doctor Nicholas Copernicus, whom I truly cherish not otherwise than as a brother. For many years I have loved him most cordially, my good will for him being more easily felt than I can express. He is renowned and recognized far and wide, not only with distinction but also with admiration in many fields of fine writing. In his old age, almost at the end of his allotted time, he is still said to let his mistress in frequently in secret assignations.

Your Reverence would perform a great act of piety if you warned the fellow privately and in the friendliest terms to stop this disgraceful behavior, and no longer let himself be led astray by Alexander [Scultetus], whom he declares to be all by himself outstanding in all respects among all our brothers, the officials and canons [of the Varmia Chapter]. If your Reverence convinces him, you will accomplish a result so pleasing to me that nothing could delight me more. In this way we shall both win over so highly esteemed a brother.

Your Reverence will conduct the conversation with him about these matters in such a way that he will recognize it is more important for him that it arose, not at my suggestion, but out of your Most Reverend Lordship's good will for him.

5 July [1539]

John [Dantiscus]
bishop of Varmia[25]

Reading No. 27

Giese Prefers to Have his Warning to Copernicus Understood as Originating with Dantiscus

Giese replied quite promptly (7 July 1539) to Dantiscus' proposal (5 July 1539) that Copernicus should be cautioned regarding his conduct. But Giese believed that Copernicus would be more deeply impressed if he understood that the prime mover in this situation was Dantiscus, with Giese acting only as the intermediary.

Most Reverend Father in Christ,
most honorable brother,
greetings and well deserved devotion:

I thank your Reverence for communicating to me here what you have heard from the [royal] court. The sort of thing that is alleged against us in the senate should, in my opinion, be credited to us. We shall have occasion to talk about it at another time.

With regard to your Reverence's warning to Doctor Nicholas [Copernicus], I shall pass it on with all my heart. But in my judgment it will have a greater influence in affecting him if he understands that I am acting at the suggestion of your Reverence. Thus he may see that he is being advised in a good and sincere spirit with a view to his reputation and standing, and may reject contact with those who perhaps persuaded him otherwise....

Lubawa, 7 July 1539[26]

Reading No. 28

Copernicus Denies Seeing Anna Schilling after her Dismissal

An event that changed the course of Copernicus' life and helped to initiate modern astronomy was Rheticus' arrival in Frombork. Rheticus had been granted a leave of absence by the University of Wittenberg, where he taught (the old) astronomy. Having heard about

Copernicus, he made up his mind to learn about the new astronomy from the master himself. On 14 May 1539 he reached Poznań (Posen, in German) where he wrote a (lost) letter to a German astronomer. When Rheticus arrived in Frombork, he was given a hearty welcome. Yet on 21 March 1539 Dantiscus had ordered all sympathizers with Lutheranism to leave Varmia within a month and never return. Rheticus came from Wittenberg, the intellectual center of Lutheranism. This confessional difference between the visitor and his host disturbed neither of them. Rheticus spent nearly ten weeks studying the lengthy manuscript which Copernicus had written with his own hand. But when Rheticus developed a slight illness, he and Copernicus were invited by Giese to Lubawa, where they spent several weeks. Using this opportunity to present Dantiscus' charges to Copernicus, Giese reported the outcome to Dantiscus on 12 September 1539.

...I have talked earnestly to Doctor Nicholas [Copernicus] about the subjects specified in your Reverence's warning. I put the situation, just as it is, before his eyes. He seemed to be disturbed not a little. For, although he has always obeyed your Reverence's wishes without delay, he is still falsely accused by malicious persons of secret assignations etc. For he denies having seen her [Anna Schilling] since her dismissal, except that she spoke to him in passing as she was leaving [Frombork] for the fair in Koenigsberg. I was absolutely convinced that he is not as emotionally involved as most people think. My conclusion is fortified also by his advanced age and unremitting studies as well as by his uprightness and honesty as a man. Nevertheless I cautioned him not to present even the appearance of misconduct. I believe he will do so. On the other hand, I think it is right for your Reverence not to have too much faith in an informer by bearing in mind that those in high places are the target of a spontaneous envy that is not afraid to molest your Reverence too. I commend myself etc.

Lubawa, 12 September 1539[27]

Reading No. 29

Dantiscus Receives a Report from the Administrator of the Varmia Chapter about Copernicus and the Other Two Canons under Attack

Among the four canons listed on the panel approved by the king of Poland on 4 September 1537 to fill the vacancy in the bishopric of Varmia was Achatius von der Trenck. As Administrator of the Varmia Chapter, Trenck was stationed in Olsztyn. He had promised to visit Bishop Giese of Chełmno. Hearing that Copernicus was Giese's guest in Lubawa, Trenck kept his promise to the bishop of Chełmno. Knowing that Giese had raised the question of Anna Schilling with Copernicus, Trenck also discussed her with the astronomer. Upon his return to Olsztyn, on 13 September 1539 Trenck sent Dantiscus an account of his conversation with Copernicus in Lubawa.

...In the proceeding against Alexander [Scultetus] in my opinion your Paternal Reverence will be able to settle hardly anything definite at the present time, since the housekeeper has not been seen in Frombork since the day she left. I visited the Reverend Bishop of Chełmno [Giese], as I had promised him a long time ago, and consulted with him how the Dean [of the Varmia Chapter, Leonard] Nidderhoff could be persuaded....But it is useless talking to him because he always sings the same song.

When Doctor Nicholas [Copernicus], whom I found at Lubawa, heard his housekeeper mentioned, he declared that he would never receive her in his house nor do anything further in this case. I know that he was advised by the Reverend Bishop of Chełmno to do so, I hope not in vain, since his age and prudence can readily deter an upright, good man from actions of this kind in the future.
Olsztyn, 13 September 1539[28]

Reading No. 30

Bishop Dantiscus' Instructions to the Varmia Chapter concerning Protestants and Prostitutes

*Bishop Dantiscus of Varmia issued an Edict against Heresy (**Mandatum wider die Ketzerey**) on 21 March 1539. Then on 15 April 1540 he repeated in Varmia the Polish king's order that the nobility's sons and intended heirs must be recalled "from the poisoned places of heretical Lutheranism" within eight months. Otherwise, their property would be confiscated, and they themselves banned forever from the realm. The same punishment was threatened against those who failed to destroy Lutheran books, songs, "or whatever has come out of the poisonous places." Making no specific reference to Rheticus, who had come out of Wittenberg, that most "poisonous" of Lutheran places, Dantiscus upbraided the Varmia Chapter for not taking parallel measures in its own jurisdiction. He also lashed out at the Chapter's inactivity in the matter of the canons' housekeepers. But if the Chapter assisted the Counter-Reformation, he would intensify the king's good will toward the Varmia Chapter.*

Instructions to Be Communicated to the Venerable Varmia Chapter [at its General Meeting about 1 November 1540 through Provost Płotowski]

Your Lordships are not unaware of what zealous and elaborate efforts have always been made by the Most Reverend John [Dantiscus], bishop of this Varmia diocese, from the very beginning of his entry into this bishopric and from the time he gained control of it, to hold this church of his unharmed by the errors of these times. He also [sought] to keep it unsullied by those [tendencies] on account of which evils are undertaken not only by those who live in this area but also among the nobles of the realm and even in the sacred court of His Majesty, our most clement lord. He [endeavored] also to preserve the diocese in a peaceful state, for our Christian glory, and the honor of his Most Reverend Lordship, and your Lordships. The venerable provost Paul Płotowski has also previously called this to the attention of your Lordships.

As a result his Most Reverend Lordship has acted against certain prostitutes here in Varmia, to have them expelled on account of their notorious life and intercourse. Their conduct has defiled this diocese of Varmia and brought upon your Lordships ecclesiastical censure even as far as the interdict and the invocation of the secular power. Yet heretofore the effort is in vain, and almost nothing has been accomplished.

Moreover, his Reverence is extremely astonished that your Lordships could heretofore, to the disgrace of the church and the entire clergy, and in disregard of the censures, tolerate and view with undisturbed eyes whores of this kind, excommunicated, denounced, blacklisted over and over again, indeed even forbidden to associate with people.

Accordingly, unless your Lordships act with greater determination in this matter of the whores, and without further delay banish them, never to return, his Reverence will be unable to forbear from proceeding to the ultimate censure (which is in his hands), and to impose an interdict here on his church, and halt services where he would most particularly want them to be performed and extended from day to day. Therefore his Reverence still urges and requests your Lordships to go far enough along in this direction so that he may not be driven [uncertain reading] and be forced to invoke this basis of the church's statutes, something he would want least of all.

Furthermore, it is not unknown to your Lordships how widely the Lutheran heresy has spread and how deeply it has sunk its roots, so that perhaps it has not failed, or does not fail, to affect the common people, that is, those on the bishop's and the Chapter's lands. Consequently his Reverence, as an obligation of his pastoral office, last year [on 21 March 1539] in a public edict ordering his subjects not to be taken in by these tricks of the heretics or corrupted by apocryphal writings, thought it worthwhile to admonish and scare them, partly by fatherly warnings, partly by quite severe threats.

As a further result it came to pass that His Sacred Royal Majesty, together with the entire senate of the notables of the realm, at the recent session in Cracow decided by unanimous voice and vote to wipe out this most harmful sect, and in addition distributed a public edict throughout the domains of the whole kingdom, and sent it to each and every one of the sections of these lands. One of the copies was received by his Reverence.

From the transcription of it which was sent to your Lordships, you have certainly understood in its entirety what His Royal Majesty desires regarding the recall of children and students living in dangerous schools and localities, infected by this heretical stain.

Aroused by the strictness of this royal edict, his Reverence again ordered [on 15 April 1540] all the classes and ranks of his subjects to submit to the king's wishes and exhibit compliance, if they wanted to avoid the punishments spelled out in the edict.

His Reverence has also deemed it worthwhile to urge your Lordships to keep the royal edict before your eyes, to conform to the kingdom of Poland, his Reverence, and all the classes of these lands, and to publish a similar edict throughout the Chapter's domains. His Reverence cannot refrain from marveling that no such step has been taken up to the present time.

Accordingly his Reverence still paternally urges your Lordships to reply in writing now through me [Provost Paul Płotowski] about anything you have done or are deciding to do about this matter in this general [meeting of the Varmia] Chapter [around 1 November 1540]. Hence, if you are going to conform to the other classes of the kingdom and these lands, and really obey the king's command, with so much greater propriety his Reverence could mention to His Royal Majesty the obedience of your Lordships, and intensify His Majesty's good will and attitude favorable to the preservation of your church's rights, privileges, and immunities. For in whatever ways his Reverence can honor and serve his church and your Reverences, he promises that his efforts in this direction will be untiring.[29]

Reading No. 31

An Attempt to Disqualify Copernicus as a Judge

Alexander Scultetus, a Varmia canon under attack at the same time as Copernicus, was ordered by the king of Poland to stand trial. Scultetus knew that the outcome of his trial in Poland was a foregone conclusion: he would be convicted. Hoping to do better in Rome, he slipped away from Frombork, leaving his canon's house empty.

It was promptly occupied by a Polish nobleman, Nicholas

Płotowski, who was sued by a counter claimant. The case was tried before the Varmia Chapter. Płotowski tried to have Copernicus declared ineligible to be a judge, on the ground that he was under suspicion, like Scultetus.

The following reports of two proceedings conducted by the Varmia Chapter are preserved somewhat imperfectly.

In the year [15]40 on the tenth [day of January] Nicholas Płotowski presented a letter from the most serene lord Sigismund, king of Poland, as well as a letter from our Most Reverend Bishop of Varmia addressed to the venerable Chapter. In offering these documents, as a subject of the crown [Płotowski argued] further how improperly and unlawfully he was summoned (the jurisdiction of the venerable Chapter was not applicable to him). Nevertheless he appeared, not because he was required to do so, but for the sake of preserving [his rights]. He therefore requested that the trial regarding the house formerly occupied by the venerable Alexander [Scultetus] (the subject matter about which the case has been brought) should be set aside, and the disposition of the entire affair referred to His Royal Majesty etc.

On the opposing side, Henry Braun, speaking through his lawyer, [argued that the case] had been remanded not merely by His Royal Majesty to the local government, but also by the Councillors of these lands to the venerable Chapter, and the venerable Chapter approved the day and the deadline established for them.

At the deadline specified in the summons the said Nicholas Płotowski, being legally summoned, on his side asked that he should be relieved of the complaints against him. [He advanced] groundless objections that the case should not be continued any longer. For his part Nicholas Płotowski contended that, according to his previous objection, he was summoned by a local judge without authority over him, and that he was not required to appear. He added, moreover, that he was unwilling to go forward with the trial of the case unless the venerable dean Leonard Nidderhoff, and Doctor Nicholas [Copernicus] withdrew from the hearing (these men being under suspicion in the same case as Scultetus, for reasons to be put forward and proved by Płotowski at an appropriate time and place). Unless Nidderhoff and Copernicus left the judges' bench to the remaining canons, Płotowski would testify at another time about his grievance as the basis for an appeal.

After these motions had been argued in various ways, in the end the venerable Chapter ruled that jurisdiction over this case clearly and lawfully belonged to themselves as the natural lords of the area. Moreover, Nicholas Płotowski (since the question concerns an area and house under the jurisdiction of the Chapter) was deemed to have been summoned lawfully and rightfully. Lastly Płotowski seemed to put forward an objection placing the venerable dean Leonard N[idderhoff] and Doctor Nicholas Copernicus under suspicion, but seemed not to produce reasons or prove them. The venerable Chapter postponed the entire case until the next St. George's day [23 April],so that then [Płotowski might bring forward] legal reasons, if he has any, why those same men [Copernicus and Nidderhoff] must not participate in the judgment. If he produces reasons, the venerable Chapter will examine them and decide whether they are rational grounds for the conclusion that the said persons [Copernicus and Nidderhoff] should be excluded from the judgment.

When the Chapter assembled to consider the case, it excluded the litigants, and ordered

them to be told that the number of judges would not be more than three. Secondly, the noble lord Nicholas Płotowski claimed at the previous session, that is, the one on 10 January, that he would advance legal reasons why the venerable dean and Doctor Nicholas [Copernicus] neither should nor could participate in deciding the case. Therefore, of necessity and by law it was requisite and suitable for them to be present in person as those against whom objections were raised. Otherwise, when an objection is made behind a person's back, there is in fact no objection. In view of the absence of each of the said persons, the venerable Chapter therefore postponed the entire case for the aforesaid reasons until the second day after the next John the Baptist day [24 June]. The action was taken as indicated above.[30]

Reading No. 32

Copernicus' Last Illness

While Copernicus' health was good, he attended conscientiously to his duties as a canon, but devoted as much time as he could to his research. This was not a team effort, as it often is today. Copernicus "loved privacy," as his best friend Giese, who lived far from Frombork, recognized. Hence, when Copernicus fell seriously ill as he was nearing his seventieth birthday, it was George Donner, a recently admitted canon, of advanced age himself, who looked after the sick old man. Donner notified Giese of Copernicus' condition, and Giese replied from Lubawa on 8 December 1542.

To George Donner:

I was shocked by what you wrote about the impaired health of the venerable old man, our Copernicus. Just as he loved privacy while his constitution was sound, so, I think, now that he is sick, there are few friends who are affected by his condition. Yet we are all indebted to him for his uprightness and outstanding learning. I know, however, that he always reckoned you among those most faithful [to him]. I therefore ask you, if his condition so warrants, please to watch over him and take care of the man whom you cherished at all times together with me. Let him not be deprived of brotherly help in this emergency. Let us not be considered ungrateful toward this deserving man. Farewell.

Lubawa, 8 December 1542[31]

Reading No. 33

Copernicus Remembers a Niece in his Will

Although Copernicus' will has not survived, something is known about a beneficiary. For he bequeathed some of his assets to a niece, who had married a military drummer in the service of the duke of Prussia. The husband asked the duke to facilitate the transmission of the assets. Accordingly, the duke addressed a letter, jointly and severally, to the four executors of Copernicus' will. The niece carried the duke's letter personally to Frombork.

To Theodoric of Reden, Leonard Nidderhoff,
George Donner, and Michael Leutze,
jointly and severally, 29 June [1543]:

First of all, my greetings and good will to you, especially beloved, worthy, honorable, learned, and upright men. I take this occasion kindly to inform you hereby that my servant and drummer, Caspar Stulpawitz, has humbly given me to understand that in his last will Nicholas Copernicus, canon of Frombork, of blessed memory, is said to have mentioned Caspar's wife Christina, the bearer of this letter, as a closely related friend, and also to have bequeathed some of his heritable property to her. For her sake, Caspar has humbly besought and begged me for a gracious letter to you, which he supposes will be of no little value to him. Without my saying so, I know that each and every one of you is disposed to carry out the stipulations of the will fairly. Yet I did not know how to resist him with propriety when he made this reasonable request on behalf of his wife. Accordingly I kindly wish you for my sake graciously to let the wife of my aforementioned servant acquire what Nicholas Copernicus bequeathed to her by means of his testament. For the purpose of having her declare this writing of mine intended for her, I am agreeable to having you jointly and severally disposed in all graciousness to so recognize it. Then my servant, Caspar Stulpawitz, often mentioned herein, will also not fail to be especially grateful to each and every one of you.[32]

Reading No. 34

The Death of Copernicus

After Copernicus' memory and mental alertness began to fail him, he suffered a hemorrhage, followed by a paralysis of the right side, with death occurring on 24 May 1543. That was the day on which Copernicus' ***Revolutions*** *in printed form reached its author, who saw it just before he closed his eyes for the last time. This account of Copernicus' death was contained in a letter sent by Giese from Lubawa on 26 July 1543 to Rheticus.*

To Joachim Rheticus:

On my return from the royal wedding in Cracow [of Prince Sigismund Augustus of Poland with Elisabeth, archduchess of Austria], in Lubawa I found the two copies, which you had sent, of the recently printed treatise of our Copernicus. I had not heard about his death before I reached Prussia. I could have balanced out my grief at the loss of that very great man, our brother, by reading his book, which seemed to bring him back to life for me. However, at the very threshold I perceived the bad faith and, as you correctly label it, the wickedness of Petreius [the Nuremberg printer-publisher of the *Revolutions*], which produced in me an indignation more intense than my previous sorrow. For who will not be anguished by so disgraceful an act, committed under the cover of good faith?

Yet I am not sure whether [this misconduct] should be attributed to this printer, who depends on the labor of others, rather than to some jealous person. Grieving that he would have to abandon the previous beliefs if this book achieved fame, perhaps he took advantage of that [printer's] ingenuousness to diminish faith in the treatise. However, lest the man should escape scot-free who permitted himself to be misled by someone else's deception, I have written to the City Council of Nuremberg, indicating what I thought had to be done in order to restore faith in the author. I am sending you the letter together with a copy of it, to enable you to decide how the affair should be managed on the basis of what has been started. For I see nobody better equipped or more eager than you to take this matter up with that City Council. It was you who played the leading part in the enactment of the drama, so that now the author's interest seems to be no greater than yours in the restoration of that which has been distorted. Provided that this interests you at all, I ardently implore you to pursue this matter with the utmost earnestness. If the first sheets are going to be printed again, it seems that you should add a brief introduction which would cleanse the stain of chicanery also from those copies which have already been distributed.

I should like in the front matter also the biography of the author tastefully written by you, which I once read. I believe that your narrative lacks nothing but his death. This was caused by a hemorrhage and subsequent paralysis of the right side on 24 May, his memory and mental alertness having been lost many days before. He saw his treatise only at his last breath on his dying day.

The distribution of the published treatise before his death will not be an obstacle, since the year agrees, and the day when the printing was finished was not indicated by the publisher. I should like also the addition of your little tract, in which you entirely correctly defended the earth's motion from being in conflict with the Holy Scriptures. In this way you will fill the volume out to a proper size and you will also repair the injury that your teacher failed to mention you in his Preface to the treatise. I explain this oversight not by his disrespect for you, but by a certain apathy and indifference (he was inattentive to everything which was nonscientific) especially when he began to grow weak. I am not unaware how much he used to value your activity and eagerness in helping him.

As for the copies of the treatise which you sent to me, I am deeply grateful to the donor. These copies will serve me as a permanent reminder to preserve the memory not only of the author, whom I always cherished, but also of you. Just as you proved yourself to be an energetic assistant to him in his labors, so now you have helped us with your effort and care lest we be deprived of the enjoyment of the finished work. It is no secret how much we all owe you for this zeal. Please let me know whether the book has been sent to the pope; for if this was not done, I would like to carry out this obligation for the deceased. Farewell.[33]

Reading No. 35

The Varmia Chapter Asks Bishop Dantiscus to Decide whether Anna Schilling Can Legally Be Banned from Frombork after Copernicus' Death

Anna Schilling, Copernicus' former housekeeper, had been ordered to leave the Varmia diocese after a trial held in the court of Bishop Dantiscus. However, she owned a house in Frombork, which she wanted to sell. Copernicus, the cause of her exclusion, was dead. In the light of the legal rule which held that when the cause ceases, the effect ceases, the Varmia Chapter could not decide whether Anna Schilling could lawfully be excluded. On 10 September 1543 they referred the question to Bishop Dantiscus, who had presided over the court of first instance.

To the Reverend John [Dantiscus] etc.

It is not unknown under what circumstances Anna Schilling was banished from here. She used to be the housekeeper of the venerable Doctor Nicholas [Copernicus], while he was alive [he had died on 24 May 1543]. Now from time to time she seems to come here and stay several days for the sake of taking care of her property, as she is said to claim. For she still owns a house here, which she is said to have sold yesterday.

We are not sure whether she will be able lawfully to be prohibited from coming here, since the legal obstacle is inoperative. For, when the cause is removed, so is the effect. Nevertheless, we are unwilling to reach any decision in this matter, unless your Paternal Reverence hands down a prior ruling on this question, since this case was first tried in your court. It will be no trouble to inform us about the decision of your Paternal Reverence, whom we commend to God.
Frombork, 10 September 1543

Officers, canons, and Chapter
of the Varmia diocese[34]

Reading No. 36

Bishop Dantiscus Banishes Anna Schilling

No sooner had Bishop Dantiscus received the Varmia Chapter's question whether Anna Schilling could lawfully be excluded from Frombork, than he curtly answered "Yes!" Since she trapped Copernicus, he argued, she could trap anyone of you, and the old scandal would revive instead of dying out.

John [Dantiscus],
by the grace of God,
bishop of Varmia
Venerable, dearly beloved, brothers:

She [Anna Schilling], who has been banned from our domain, has betaken herself to you, my brothers. I am not much in favor, whatever the reasons. For it must be feared that by the methods by which she deranged him [Copernicus], who departed from the living a short while ago, she may take hold of another one of you, my brothers. But if you have decided to let her stay among your people, that is to be judged by you, my brothers. Nevertheless, I would consider it better to keep at a rather great distance than to let in the contagion of such a disease. How much she has harmed our church is not unknown to you, my brothers, for whom I hope happiness and health.

In my palace in Lidzbark
13 September 1543[35]

Reading No. 37

"Gresham's Law" Proclaimed, and Later Repealed

"Bad money always drives good money out from circulation." This was how Henry Dunning MacLeod (1821-1902) formulated the principle which he dubbed "Gresham's Law," attributing it to Sir Thomas Gresham (c. 1519-1579).

Gresham...has the great merit of being as far as we can discover, the first who discerned the great fundamental law of the currency, that good and bad money cannot circulate together. The fact had been repeatedly observed before, as we have seen, but no one, that we are aware, had discovered the necessary relation between the facts, before Sir Thomas Gresham....He was presented to the Queen only three days after her accession...and she immediately employed him to negociate *[sic]* a loan which was necessary in the exhausted state of the Treasury....Before leaving for Flanders, he wrote a letter of advice to the queen explaining how, among other things, all the fine money had disappeared from circulation. The cause of this he attributed *to the debasing of the coinage by Henry VIII.* [MacLeod's emphasis] Now, as he was the first to perceive that a bad and debased currency is the *cause* [MacLeod's emphasis] of the disappearance of the good money, we are only doing what is just, in calling this great fundamental law of the currency by his name. We may call it Gresham's law of the currency.[36]

Nearly forty years later, in his ***Law of Gresham and its Relation to Bimetalism*** *(London, 1896), p. 7, MacLeod referred to "Gresham's Law, which name has now been universally accepted." Although Gresham's letter to Queen Elizabeth I was cited by MacLeod, he failed to identify any expression of "Gresham's Law" in it. "Nowhere does Gresham state either explicitly or implicitly that bad money drives out the good."[37] In fact, Gresham noticed that of two debased coinages, the better continued to circulate when issued in limited quantity, not in excess of the needs of trade.[38] Many people before Copernicus had understood what was later miscalled "Gresham's Law." But its appearance in a monetary treatise earlier than Copernicus' has not yet been documented.*

Reading No. 38

Peurbach's Legacy to Regiomontanus

For university students of astronomy in the fifteenth century, Ptolemy's ***Syntaxis*** *should have been the standard reference work. But Greek, the language in which the* ***Syntaxis*** *was written, was not being* *taught. A Latin translation had been made in the twelfth century. However, it had not been based on the original Greek text. Instead, it relied on an Arabic version of the Greek. As a result, this medieval Latin translation was most unsatisfactory.*

Professor Peurbach of the University of Vienna was advised by Cardinal Bessarion, a vigorous proponent of Greek studies, "to try to make Ptolemy briefer and clearer" in a Latin version. Peurbach's **Epitome** *reached only as far as Book VI (of the thirteen Books in the* **Syntaxis**) *when he lay dying in the arms of his favorite pupil, Johannes Regiomontanus (1436-1476). Peurbach's last words to his disciple were as follows.*

Farewell, he said, my dear Johannes, farewell. And if the remembrance of your faithful teacher can affect you at all, finish the work of Ptolemy which I am leaving unfinished. This I bequeath to you as a legacy, so that even though I have departed this life, with the better part of me surviving I may carry out the wishes of our most excellent and most honorable leader, Bessarion.[39]

Reading No. 39

Why Regiomontanus Settled in Nuremberg

After the years Regiomontanus had spent at the University of Vienna, in Bessarion's circle, and in the service of the king of Hungary, he decided to settle in Nuremberg. He explained his reasons for doing so in a letter to a fellow mathematician, the rector of the University of Erfurt.

I have chosen Nuremberg as my permanent home not only on account of the availability of instruments, particularly the astronomical instuments on which the entire science of the heavens is based, but also on account of the very great ease of all sorts of communication with learned men living everywhere, since this place is regarded as virtually the center of Europe because of the journeys of the merchants.[40]

Reading No. 40

The Publication of the *Epitome*

The **Epitome** *was dedicated to Bessarion by Regiomontanus. He had a professional scribe prepare a magnificent copy of the work on parchment. It was presented to the cardinal before the end of April*

1463. Throughout the remaining years of his life Bessarion declined to have the **Epitome** *printed.*

Regiomontanus had kept his own copy. When he became the first publisher of mathematical and astronomical books, he announced his intention to print the **Epitome.** *But his untimely death prevented him from doing so.*

Bessarion's copy of the **Epitome** *languished in the library of St. Mark in Venice until it was noticed by an active editor of astronomical works. In 1492 he made public his wish to print the* **Epitome.** *But he never did. On the other hand, another editor applied to the Venetian authorities for the customary copyright of ten years. In his application, which was filed on 10 February 1496, he gave the following explanation.*

I have expended a great deal of time and effort as well as very large sums in locating, correcting, and preparing my own diagrams for an astronomical work entitled the *Epitome* of Johannes Regiomontanus, a most accomplished scholar, very proficient in the science of astronomy. This work has never been printed, for it is rare and also has been seen by very few scholars. This is because everybody who could get hold of it kept it hidden as his own treasure to prevent other specialists from asking to borrow it.[41]

The petitioner's request was approved by the Venetian authorities. On 31 August 1496 the **Epitome** *was finally published. The presentation was worthy of the contents: it is an imposing and splendid volume, folio size, a beautiful example of early Venetian printing at its best. It called Bessarion "Patriarch of Constantinople." This is a title he had not yet received when the parchment copy of the* **Epitome** *was presented to him before the end of April 1463.*

Reading No. 41

The *Epitome* Called Attention to a Flaw in Ptolemy's Astronomy

The title of the **Epitome** *suggests that it was nothing more than a summary of Ptolemy's* **Syntaxis,** *a long and involved treatise. But the* **Epitome** *is more than a summary. As the moon revolves around the earth once a month, according to Ptolemy's lunar theory sometimes it is almost twice as far away as at other times. If so, its diameter would have to look twice as long at one time in a month as at another time. Actually, there is a variation. But it is much smaller than 2:1. Ptolemy himself gave no hint that his lunar theory departed so far from nature.*

*The **Epitome**, however, pounced upon this discrepancy between Ptolemy and the observed facts. Ptolemy's lunar theory combines two major components: an epicycle riding on an eccentric deferent. The moon is at its closest to the earth when it is in the perigee of the epicycle, while the epicycle's center is in the perigee of the eccentric deferent.*

The moon attains its least distance [from the earth], that is, 33 1/2 earth-radii, when it is opposite the apogee of the eccentric and epicycle....But it is noteworthy that when the moon is opposite the epicycle's apogee at half-moon, it does not appear so big. Yet if its entire disk [rather than half of it] were visible, it would have to look four times the size which appears at opposition [full moon], when it is in the epicycle's apogee.[42]

Reading No. 42

Copernicus' Rejection of the Equant

*Copernicus' **Commentariolus** objected to Ptolemy's lunar theory because it conflicted with the observations by making the moon look bigger than it does in nature. This is an objection based on the facts. The **Commentariolus** also had a theoretical objection to Ptolemy's theory of the moon: it used nonuniform motion. In this arrangement, the motion of a point traveling along the circumference of a circle is nonuniform as seen from the circle's center, from which the moving point's distance is constant. The point's motion is uniform, however, as seen from a different point outside the center of the circle. But the distance of the moving point from this eccenter is not constant.*

The widespread [planetary theories], advanced by Ptolemy and most other [astronomers], although consistent with the numerical [data], seemed likewise to present no small difficulty. For these theories were not adequate unless they also conceived certain equalizing circles, which made the planet appear to move at all times with uniform velocity neither on its deferent sphere nor about its own [epicycle's] center. Hence this sort of notion seemed neither sufficiently absolute nor sufficiently pleasing to the mind.[43]

What sufficiently pleased Copernicus' mind was absolute uniform motion. This required a point traveling along the circumference of a circle to maintain a uniform distance from the center of the circle. At the same time the speed of the moving point must be uniform as seen from the center of the circle.

Reading No. 43

An Earlier Rejection of Ptolemy's Departure from Absolute Uniform Motion

The eminent Islamic scientist Ibn al-Haytham (965-c.1040) wrote a commentary on Ptolemy's **Syntaxis**, *a cosmological treatise in the Ptolemaic tradition, and a study of the moon's motion. In none of these three works did he say a word in opposition to Ptolemy. However, in a (lost) treatise on Ptolemy's theory of planetary latitudes, he indicated dissatisfaction with the Greek astronomer's discussion. For this reason he was rebuked by a correspondent, who thought that Ptolemy's authority should not be questioned. Against this traditionalist point of view, Ibn al-Haytham composed his* **Doubts (Shukuk) about Ptolemy**, *which is critical of Ptolemy's* **Syntaxis**, **Planetary Hypotheses**, *and* **Optics**. *In the* **Shukuk**'s *section dealing with the* **Syntaxis**, *Ibn al-Haytham included the following remarks.*

Ptolemy assumed an arrangement that cannot exist, and the fact that this arrangement produces in his imagination the motions that belong to the planets does not free him from the error he committed in his assumed arrangement, for the existing motions of the planets cannot be the result of an arrangement that is impossible to exist....

The way Ptolemy followed was indeed a legitimate beginning, but since it led him to what he himself admitted to be contrary to the rules, he should have declared his assumed arrangement to be false....

Ptolemy gathered all that his predecessors and he himself had verified regarding the motions of each one of the planets. Then he searched for an arrangement that should belong to existing bodies having these motions. Having failed in his search, he assumed an imaginary arrangement in terms of imaginary circles and lines to which these motions are attributed....But for a man to imagine a circle in the heavens and to imagine the planet moving in it does not bring about the planet's motion....And therefore the arrangements assumed by Ptolemy for the five planets are false, and he asserted them knowing them to be false, and there exists for the planets a true arrangement in existing bodies which Ptolemy failed to grasp. For there cannot be a sensible, permanent, and ordered motion unless it has a true arrangement in existing bodies.[44]

Reading No. 44

The Alleged Retrogression of the Solar Apogee

In the Ptolemaic theory, the sun performed an annual revolution around the earth. The center of the sun's circle lay outside the earth. Hence, the sun reached a point, its apogee, where it was farthest from the earth. Ptolemy located this apogee precisely where his great predecessor Hipparchus had placed it. Accordingly, Ptolemy concluded that the position of the solar apogee was fixed. Later astronomers, however, found that the solar apogee lay to the east of Ptolemy's position. Thus, the ***Epitome*** *reported that al-Battani (called Albategnius in Latin; +929) had put the solar apogee 16°47′ farther east than Ptolemy (16°47′ + 65°30′ = 82°17′ east of the vernal equinox = 7°43′ west of the summer solstice). But the* ***Epitome*** *also asserted that the solar apogee's position, as determined by al-Zarqali (or Arzachel in Latin; +1100), was 4°27′ west of al-Battani's (4°27′ + 7°43′ = 12°10′ west of the summer solstice.)*

The arc AB is the distance from the solar apogee to the first point of the Ram [= vernal equinox], a distance which Ptolemy found to be 65½°, just as it had been found by Hipparchus. Hence Ptolemy concluded that the sun's apogee is immovable and fixed in relation to the vernal and autumnal equinoxes. Al-Battani found...that the arc BH [between the solar apogee and the summer solstice] was 7°43′....Al-Zarqali, however,...set the arc BH = 12°10′. This certainly looks strange, since al-Zarqali lived after al-Battani, whose observation you may trust....Al-Zarqali, 193 years after al-Battani,...finding BH = 12°10′, was therefore compelled to say that the center of the sun's eccentric moved on a certain small circle.[45]

Reading No. 45

Copernicus Overcomes an Error in the *Epitome*

The ***Epitome*** *reported that al-Zarqali found the solar apogee to the west of al-Battani's determination. Yet al-Zarqali lived nearly two centuries later than al-Battani. "This certainly looks strange," commented the* ***Epitome****, adding that "you may trust al-Battani's observation." Without saying so explicitly, the* ***Epitome*** *cast doubt on al-Zarqali.*

Like the ***Epitome's*** *authors, Peurbach and Regiomontanus, Copernicus did not know Arabic. Hence, he was in no position to decide, as between al-Battani and al-Zarqali, who was right and who was wrong. Both of them enjoyed the highest reputation as* *astronomers. Yet, something was obviously amiss. Copernicus accepted the* ***Epitome****'s reports about al-Battani's and al-Zarqali's determinations of the solar apogee as equally valid. But he discarded al-Zarqali's. For in the long history of the determinations of the solar apogee, al-Zarqali's was the only report of a backward or retrogressive apogeal motion.*

The shift in the solar apse now presents a problem which is more acute because, whereas Ptolemy regarded the apse as fixed, others thought that it accompanied the motion of the sphere of the stars, in conformity with their doctrine that the fixed stars move too. Al-Zarqali believed in the nonuniformity of this [motion], which even happened to regress. He relied on the following evidence. Al-Battani had found the apogee...7°43′ ahead of the solstice: in the 740 years since Ptolemy it had advanced nearly 17° ($\cong 24^\circ 30' - 7^\circ 43'$). In the 193 years thereafter it seemed to al-Zarqali to have retrogressed about 4½° ($\cong 12^\circ 10' - 7^\circ 43'$). He therefore believed that the center of the annual orbit had an additional motion on a circlet. As a result, the apogee was deflected back and forth....

[Al-Zarqali's] idea was quite ingenious, but it has not been accepted because it is inconsistent with the other findings taken as a whole. Thus, consider the successive stages of that motion. For some time before Ptolemy it stood still. In 740 years or thereabouts it progressed through 17°. Then in 200 years it retrogressed 4° or 5°. Thereafter until our age it moved forward. The entire period has witnessed no other retrogression nor the several stationary points which must intervene at both limits when motions reverse their direction. [The absence of] these [retrogressions and stationary points] cannot possibly be understood in a regular and circular motion. Therefore many [specialists] believe that some error occurred in the observations of those [astronomers, that is, al-Battani and al-Zarqali]. Both [were] equally skillful and careful practitioners so that it is doubtful which one we should prefer to follow.

For my part I confess that nowhere is there a greater difficulty than in understanding the solar apogee....It is highly probable, therefore, as we can deduce from the general structure of the motion, that it is direct, yet nonuniform. For after that stationary [interval] from Hipparchus to Ptolemy the apogee appeared in a continuous, regular, and progressive advance until the present time. An exception occurred between al-Battani and al-Zarqali through a mistake (it is believed), since everything else seems to fit.[46]

Reading No. 46

Al-Battani's Computation of the Position of the Solar Apogee

The Arabic text of al-Battani's masterpiece was printed toward the end of the nineteenth century. The editor then translated it into Latin. He added a useful commentary, also in Latin.

Al-Battani made a special study of the sun in a year which he numbered 1194 of the era of the Two-Horned (Dhu'l qarnayn). This was how Alexander the Great was depicted in art. Alexander's era started, according to al-Battani's reckoning, on 1 September 312 B.C. Hence, the year 1194 of Alexander's era began on 1 September 882 of the Christian era.

Al-Battani divided his great work into 57 chapters. In his solar theory, he assigned an eccentric to the sun. In the diagram accompanying Chapter 28, he called the point T on the sun's eccentric the place of the solar apogee. Then he proceeded:

It is therefore clear that the apogeal point T of the eccentric is 7°43′ away from the point of the summer solstice in the reverse order of the [zodiacal] signs, that is, 82°17′ from the first point of the Ram. This is what I wanted to find. The observation on which I based this calculation was made in the year 1194 of the era of the Two-Horned [882 A.D.], in which I observed the progress of the sun from the first point of the Ram to the first point of the Crab and the first point of the Balance.[47]

Reading No. 47

The *Commentariolus* Provided the First Correct Explanation of the Planetary Loops

As a planet moves through the heavens, its angular velocity is observed to be nonuniform. This nonuniformity used to be called its "first anomaly" or irregularity. In its "second anomaly," it slowed down, stopped, and reversed direction while deviating in latitude. The planetary loop engendered thereby was deemed to be a physical phenomenon. In a historic breakthrough, the **Commentariolus** *showed that these planetary loops are optical, not physical. For while we watch the skies, the earth moves around the Grand Orb in its annual circuit, affecting how we see a planet. An observer on the sun, regarded as stationary, would see the planet moving around him without ever stopping or changing direction.*

There is a second anomaly, in which the outer planet is seen sometimes to retrograde and often to become stationary. This second anomaly happens by reason of the motion, not of the planet, but of the earth as it changes its observational position on the Grand Orb. For since the earth's speed surpasses the motion of the planet, the line of sight directed toward the firmament regresses, and the earth more than neutralizes the planet's motion. This regression peaks at the time when the earth is nearest to the planet, that is, when it comes between the sun and the planet at the planet's evening rising. On the other hand, about the time when the planet is setting in the evening or rising in the morning, the earth advances the line of sight in the forward direction. But when the line of sight is moving in the direction opposite to the planet's and at an equal rate, the planet seems to stand still because the opposite motions neutralize each other in this way. This generally happens when the angle at the earth between the sun and the planet is about 120°. In all these planets, however, the lower the sphere by which the planet is moved, the greater is this anomaly. Hence in Saturn it is smaller than in Jupiter, and again greatest in Mars, in accordance with the ratio of the Grand Orb's radius to their radii. This anomaly peaks for each of them at the time when the planet is seen along a line of sight tangent to the Grand Orb's circumference.[48]

Reading No. 48

Apollonius of Perga's Theorems concerning the Stationary Points of the Planets

The eminent ancient Greek mathematician Apollonius of Perga wrote a treatise containing theorems about the planetary stations. The treatise itself is lost. But the theorems survive, because they were repeated by Ptolemy in the ***Syntaxis****, XII, 1. They were later transcribed from Greek into Latin by Regiomontanus in the* ***Epitome****, XII, 1-2. The* ***Epitome****, V, 22, on lunar theory, was used by Copernicus when he was writing the* ***Commentariolus****. But at that time he was unfamiliar with* ***Epitome*** *XII, 1-2. For in the* ***Commentariolus****, following Pliny the Elder's* ***Natural History****, he made the outer planets stationary when they were 120° from the sun, as seen from the earth. This primitive approximation disappeared from Copernicus' thinking after he found out about Apollonius' theorems. He used them in the* ***Revolutions****, Book V, which he wrote about a quarter of a century later than the* ***Commentariolus****.*

There are two reasons why a planet's uniform motion appears nonuniform: the earth's motion, and the planet's own motion. I shall explain each of the nonuniformities in general and separately with a visual demonstration, in order that they may be better distinguished from each other. I shall begin with the nonuniformity which is intermingled with them all on account of the earth's motion, and I shall start with Venus and Mercury, which are enclosed within the earth's circle.

Let circle AB, eccentric to the sun, be described by the earth's center in the annual revolution....Let AB's center be C. Now let us assume that the planet has no irregularity other than that which it would have if we made it concentric with AB. Let the concentric be DE, either Venus' or Mercury's. On account of their latitude DE must be inclined to AB. But for the sake of an easier explanation, let them be conceived in the same plane. Put the earth in point A, from which draw lines of sight AFL and AGM, tangent to the planet's circle at points F and G. Let ACB be a diameter common to both circles.

Let both bodies, I mean, the earth and the planet, move in the same direction, that is, eastward, but let the planet be faster than the earth. Hence C and line ACB will appear to an observer traveling with A to move with the sun's [apparent] mean motion. On the other hand, on circle DFG, as though it were an epicycle, the planet will traverse arc FDG eastward in more time than the remaining arc GEF westward. In arc FDG it will add the entire angle FAG to the sun's mean motion, while in arc GEF it will subtract the same angle. Therefore, where the planet's subtractive motion, especially near perigee E, exceeds C's additive motion, to the extent of that excess it seems to [the observer in] A to retrograde, as happens in these planets. In their cases, line CE: line AE>A's motion: planet's motion, according to the theorems of Apollonius of Perga....But where the additive motion equals the subtractive [counteracting each other], the planet will seem stationary, all these aspects being in agreement with the phenomena.

Therefore, if there were no other irregularity in the planet's motion, as Apollonius thought, these constructions could be sufficient. But these planets' greatest elongations from the sun's mean place in the mornings and evenings, as indicated by angles FAE and GAE, are not found everywhere equal. Nor is either one of these greatest elongations equal to the other, nor are their sums equal to each other. The inference is obvious that they do not move on circles concentric with the earth's, but on certain other circles by which they produce a second inequality.

The same conclusion is proved also for the three outer planets, Saturn, Jupiter, and Mars, which completely encircle the earth. Reproduce the earth's circle from the preceding diagram. Assume DE outside it and concentric with it in the same plane. On DE put the planet at any point D, from which draw straight lines DF and DG tangent to the earth's circle at points F and G, and [from D also draw] DACBE, the diameter common [to both circles]. On DE, the line of the sun's [apparent] motion, the true place of the planet, when it rises at sunset and is closest to the earth, will obviously be visible (only to an observer at A). For when the earth is at the opposite point B, although the planet is on the same line, it will not be visible, having become blotted out on account of the sun's closeness to C. But the earth's travel exceeds the planet's motion. Hence throughout the apogeal arc GBF it will appear to add the whole angle GDF to the planet's motion, and to subtract it in the remaining arc FAG, but for a shorter time, FAG being a smaller arc. Where the earth's subtractive motion exceeds the planet's additive motion (especially around A), the planet will seem to be left behind by the earth and to move westward, and to stand still where the observer sees the least difference between the opposing motions.

Thus all these phenomena, which the ancient astronomers sought [to explain] by means of an epicycle for each planet, happen on account of the single motion of the earth, as is again clear. Contrary to the view of Apollonius and the ancients, however, the planet's motion is not found uniform, as is proclaimed by the earth's irregular revolution with respect to the planet. Consequently the planets do not move on a concentric, but in another way.[49]

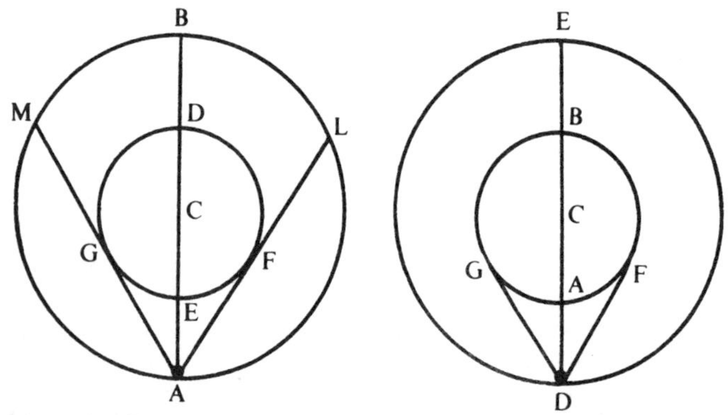

Reading No. 49

Rheticus Requests the Duke of Prussia to Ask the Elector of Saxony and the University of Wittenberg to Permit Rheticus to Publish Copernicus' *Revolutions*

Rheticus spent more than two years in Frombork and Prussia. During that stay he helped to persuade Copernicus to finish the

Revolutions and release it to the press. Copernicus finally consented to have a fair copy of his autograph manuscript made for delivery to the printer. Rheticus had devised a little instrument for determining the length of the day throughout the year. In presenting this device to the duke of Prussia, Rheticus at the same time asked him to send letters to the Elector of Saxony and the University of Wittenberg, requesting permission for Rheticus to have Copernicus' ***Revolutions*** *printed.*

...Your Highness is also willing to send a gracious request to the illustrious Elector of Saxony and the honorable University of Wittenberg that I may be permitted to give my teacher's work to the press, in accordance with what I transmitted to Your Highness through Your Highness' servant, Jerome Schürstab. I am therefore humbly thankful to Your Highness, in offering to be forever obligated with my greatest diligence in complete obedience to Your Highness. I hereby put myself humbly and loyally at Your Highness' service, as may always be graciously approved by God Almighty.
Frombork, 29 August 1541

Your Highness' willing and
humble servant
George Joachim Rheticus[50]

Reading No. 50

The Duke of Prussia Writes to the Elector of Saxony and the University of Wittenberg on Rheticus' Behalf

Before leaving Prussia, Rheticus explained his situation at the University of Wittenberg to the duke of Prussia's secretary. Rheticus was concerned whether the university would reinstate him after his return from Prussia. He was also worried whether his wages would be interrupted if he attended to the printing of Copernicus' ***Revolutions*** *outside Wittenberg. Somehow the duke's secretary misunderstood Rheticus to be talking about a book written by himself. Whether this misunderstanding affected the actual printing of the* ***Revolutions*** *is not made clear by any surviving documents.*

To the Elector of Saxony
1 September 1541

First of all, our friendly service and whatever we always possess of increasing love and good will.

Highborn prince, dear affectionate uncle and brother-in-law, the honorable and learned professor, our especially beloved George Joachim Rheticus, professor of mathematics in Wittenberg, spent some time reputably and well here in these lands of Prussia. He also pursued his science of astronomy etc. in the same manner with divine grace and help. Hence we hope that he should bring not only to Your Highness but also to the entire university no less credit, fame, and glory than, in particular, advantage and piety. We are informed by him that previously he was appointed to the professorship of astronomy in Wittenberg. He held it heretofore on the best of terms, and has not yet been validated and confirmed.

In addition, we have learned that he has put together and finished a book in his science here in these lands with great energy, effort, and labor. He thought of letting it be printed publicly abroad. We have held him in high regard on account of his scientific attainments. Hence we could not fail to write a gracious communication to Your Highness.

Accordingly, it is our friendly request to Your Highness that in recognition of his skill, ability, and value, you may wish to validate and confirm him in the aforementioned professorship which he formerly held in Wittenberg. You may also wish graciously to allow and permit him to betake himself for a time, without interruption of his professor's wages, for the sake of carrying out his intended work in the place where he decided to have his book printed. For our sake also, you may wish to show and evince to him all the gracious, beneficial good will, of which we have no doubt.

This is what we want to ask once more, with all friendly good will of a brother-in-law. We commend you herewith to God's protection and safe keeping.

Sent in similar form to the University of Wittenberg.

By direct order of the duke,
Jerome Schürstab[51]

Reading No. 51

Copernicus Is Condemned in Martin Luther's Dining Room

Luther's dining room was regularly crowded with many guests, eager to hear their leader's opinion about all sorts of subjects. When Rheticus was just beginning his momentous visit to Copernicus, the new astronomy became a topic of conversation at Luther's table. Among those present, there were some who took notes. These were published, after Luther's death, as his **Table Talk (Tischreden).** *Two slightly different versions of Luther's remarks about Copernicus have come down to us. The milder account follows.*

There was mention of a certain new astrologer who wanted to prove that the earth moves and not the sky, the sun, and the moon.

This would be as if somebody were riding on a cart or in a ship and imagined that he was standing still while the earth and the trees were moving. [Luther remarked,] "So it goes now. Whoever wants to be clever must agree with nothing that others esteem. He must do something of his own. This is what that fellow does who wishes to turn the whole of astronomy upside down. Even in these things that are thrown into disorder I believe the Holy Scriptures, for Joshua commanded the sun to stand still and not the earth."[52]

In Joannes Aurifaber's edition of the **Table Talk** *(Eisleben, 1566), #2919, Luther called Copernicus a "fool." This derisive epithet is not present in Anthony Lauterbach's version* ***(D. Martin Luthers Werke,*** *Tischreden IV, Weimar, 1916, pp.. 412-413, #4638).*

Reading No. 52

Relativity of Motion

In Luther's dining room, relative motion was urged as an argument against Copernicanism: a passenger on a moving ship imagines himself to be still, while the world outside is in motion.

This principle of relative motion lies at the very heart of Copernicus' innovatory astronomy. In **Revolutions,** *I, 5, he wrote:*

Every observed change of place is caused by a motion of either the observed object or the observer or, of course, by an unequal displacement of each. For when things move with equal speed in the same direction, the motion is not perceived, as between the observed object and the observer, I mean. It is the earth, however, from which the celestial ballet is beheld in its repeated performances before our eyes. Therefore, if any motion is ascribed to the earth, in all things outside it the same motion will appear, but in the opposite direction, as though they were moving past it. Such in particular is the daily rotation, since it seems to involve the entire universe except the earth and what is around it. However, if you grant that the heavens have no part in this motion but that the earth rotates from west to east, upon earnest consideration you will find that this is the actual situation concerning the apparent rising and setting of the sun, moon, stars and planets.[53]

Copernicus held that "the motion of the heavenly bodies is circular, since the motion appropriate to a sphere is rotation in a circle" ***(Revolutions,*** *I, 4). In I, 8, he reminded his readers that*

We regard it as a certainty that the earth, enclosed between poles, is bounded by a spherical surface. Why then do we still hesitate to grant it the motion appropriate by nature to its form rather than attribute a movement to the entire universe, whose limit is unknown and unknowable? Why should we not admit, with regard to the daily rotation, that the appearance is in the heavens and the reality in the earth? This situation closely resembles what Vergil's Aeneas says:

> Forth from the harbor we sail, and the land and the cities slip backward (***Aeneid,*** *III, 72).*

For when a ship is floating calmly along, the sailors see its motion mirrored in everything outside, while on the other hand they suppose that they are stationary, together with everything on board. In the same way, the motion of the earth can unquestionably produce the impression that the entire universe is rotating.[54]

Reading No. 53

Copernicus and the Bible

When Rheticus published his ***First Report*** *in 1540, he sent a copy to his friend Achilles Pirmin Gasser (1505-1577), together with a letter which has not come down to us. Shortly after receiving Rheticus' (lost) letter, Gasser wrote a Foreword for the* ***First Report's*** *second edition (Basel, 1541). According to Gasser's Foreword, Rheticus'* ***First Report*** *"could be considered heretical (as the monks would say)."*[55]

In September 1543 Petreius, the publisher of Copernicus' ***Revolutions,*** *presented Gasser with a copy.*[56] *On folio 2 verso of the first signature Gasser noted that Copernicus' Dedication was "composed at Frombork in Prussia in the month of June 1542."*[57] *Copernicus' Dedication was addressed to the reigning pope, Paul III, whom he asked to protect him from calumnious attacks.*

I correlate the motions of the other planets and of all the spheres with the movement of the earth so that I may thereby determine to what extent the motions and appearances of the other planets and spheres can be saved if they are correlated with the earth's motions. I have no doubt that acute and learned astronomers will agree with me, if, as this discipline especially requires, they are willing to examine and consider, not superficially but thoroughly, what I adduce in this volume in proof of these matters.

However, in order that the educated and uneducated alike may see that I do not run away from the judgment of anybody at all, I have preferred dedicating my studies to Your Holiness rather than to anyone else. For even in this very remote corner of the earth where I live you are considered the highest authority by virtue of the loftiness of your office and your love for all literature and astronomy too. Hence by your prestige and judgment you can easily suppress calumnious attacks although, as the proverb has it, there is no remedy for a backbite.

Perhaps there will be babblers who claim to be judges of astronomy although completely ignorant of the subject and, badly distorting some passage of Scripture to their purpose, will dare to find fault with my undertaking and censure it. I disregard them even to the extent of despising their criticism as unfounded. For it is not unknown that Lactantius, otherwise an illustrious writer but hardly an astronomer, speaks quite childishly about the earth's shape, when he mocks those who declared that the earth has the form of a globe. Hence scholars need not be surprised if any such persons will likewise ridicule me.

Astronomy is written for astronomers. To them my work too will seem, unless I am mistaken, to make some contribution also to the Church, at the head of which Your Holiness now stands. For not so long ago under Leo X the Lateran Council considered the problem of reforming the ecclesiastical calendar. The issue remained undecided then only because the lengths of the year and month as well as the motions of the sun and moon were regarded as not yet adequately measured. From that time on, at the suggestion of that most distinguished man, Paul, bishop of Fossombrone, who was then in charge of this matter, I have devoted my attention to a more precise study of these topics. But what I have accomplished in this regard, I leave to the judgment of Your Holiness in particular and of all other learned astronomers. And lest I appear to Your Holiness to promise more about the usefulness of this volume than I can fulfill, I now turn to the work itself.[58]

Reading No. 54

Copernicus' Reason for Delaying the Publication of the *Revolutions*

Copernicus did not completely agree with the cosmology formulated by the ancient Pythagoreans. But he did approve of their policy regarding public relations. Their finest thoughts might be reviled if released to people not properly prepared to receive them. Hence, only to kinsmen and friends should the innermost secrets be revealed. The truth about the earth - that it is a heavenly body in motion - might shock ordinary people, brought up for centuries to believe that it is a non-heavenly body at rest. Who could tell how they would react to such a disclosure? Copernicus' deep-seated reluctance to publish the **Revolutions** *was finally overcome by the insistence of a few friends.*

To his Holiness, Pope Paul III, Nicholas Copernicus' Preface to his Books on the Revolutions

I can readily imagine, Holy Father, that as soon as some people hear that in this volume, which I have written about the revolutions of the spheres of the universe, I ascribe certain motions to the terrestrial globe, they will shout that I must be immediately repudiated together with this belief. For I am not so enamored of my own opinions that I disregard what others may think of them. I am aware that a philosopher's ideas are not subject to the judgment of ordinary persons, because it is his endeavor to seek the truth in all things, to the extent permitted to human reason by God. Yet I hold that completely erroneous views should be shunned. Those who know that the consensus of many centuries has sanctioned the conception that the earth remains at rest in the middle of the heaven as its center would, I reflected, regard it as an insane pronouncement if I made the opposite assertion that the earth moves. Therefore I debated with myself for a long time whether to publish the volume which I wrote to prove the earth's motion or rather to follow the example of the Pythagoreans and certain others, who used to transmit philosophy's secrets only to kinsmen and friends, not in writing but by word of mouth....And they did so, it seems to me, not, as some suppose, because they were in some way jealous about their teachings, which would be spread around; on the contrary, they wanted the very beautiful thoughts attained by great men of deep devotion not to be ridiculed by those who are reluctant to exert themselves vigorously in any literary pursuit unless it is lucrative; or if they are stimulated to the nonacquisitive study of philosophy by the exhortation and example of others, yet because of their dullness of mind they play the same part among philosophers as drones among bees. When I weighed these considerations, the scorn which I had reason to fear on account of the novelty and unconventionality of my opinion almost induced me to abandon completely the work which I had undertaken.

But while I hesitated for a long time and even resisted, my friends drew me back. Foremost among them was the cardinal of Capua, Nicholas Schönberg, renowned in every field of learning. Next to him was a man who loves me dearly, Tiedemann Giese, bishop of Chełmno, a close student of sacred letters as well as of all good literature. For he repeatedly encouraged me and, sometimes adding reproaches, urgently requested me to publish this volume and finally permit it to appear after being buried among my papers and lying concealed not merely until the ninth year but by now the fourth period of nine years. The same conduct was recommended to me by not a few other very eminent scholars. They exhorted me no longer to refuse, on account of the fear which I felt, to make my work available for the general use of students of astronomy. The crazier my doctrine of the earth's motion now appeared to most people, the argument ran, so much the more admiration and thanks would it gain after they saw the publication of my writings dispel the fog of absurdity by most luminous proofs. Influenced therefore by these persuasive men and by this hope, in the end I allowed my friends to bring out an edition of the volume, as they had long besought me to do.[59]

Reading No. 55

Cardinal Schönberg's Letter to Copernicus

Johann Albrecht Widmanstetter (1506-1577), secretary of Pope Clement VII, "explained the Copernican opinion about the motion of the earth" to the pope, in the presence of two cardinals, a bishop, and a physician, in the Vatican gardens in the summer of 1533. The explanation pleased Clement VII so much that he presented to Widmanstetter a valuable Greek manuscript, which still survives. After Clement VII's death on 25 September 1534, Widmanstetter entered the service of Nicholas Schönberg, who later became a cardinal. Cardinal Schönberg's interest in Copernicus' astronomy was aroused by Widmanstetter to such a pitch that on 1 November 1536 the cardinal signed the following letter to Copernicus.

Nicholas Schönberg, Cardinal of Capua, to Nicholas Copernicus, Greetings.

Some years ago word reached me concerning your proficiency, of which everybody constantly spoke. At that time I began to have a very high regard for you, and also to congratulate our contemporaries among whom you enjoyed such great prestige. For I had learned that you had not merely mastered the discoveries of the ancient astronomers uncommonly well but had also formulated a new cosmology. In it you maintain that the earth moves; that the sun occupies the lowest, and thus the central, place in the universe; that the eighth heaven remains perpetually motionless and fixed; and that, together with the elements included in its sphere, the moon, situated between the heavens of Mars and Venus, revolves around the sun in the period of a year. I have also learned that you have written an exposition of this whole system of astronomy, and have computed the planetary motions and set them down in tables, to the greatest admiration of all. Therefore with the utmost earnestness I entreat you, most learned sir, unless I inconvenience you, to communicate

this discovery of yours to scholars, and at the earliest possible moment to send me your writings on the sphere of the universe together with the tables and whatever else you have that is relevant to this subject. Moreover, I have instructed Theodoric of Reden to have everything copied in your quarters at my expense and dispatched to me. If you gratify my desire in this matter, you will see that you are dealing with a man who is zealous for your reputation and eager to do justice to so fine a talent. Farewell.
Rome, 1 November 1536[60]

Reading No. 56

Tolosani's Condemnation of Copernicus' *Revolutions*

*The **Revolutions** was dedicated to Pope Paul III. Neither Copernicus nor his publisher nor his editor asked for permission to do so. After the **Revolutions** was printed, a friend of the (now deceased) author sent a copy to the pope. He turned it over to his personal theologian, Bartolomeo Spina of Pisa, whom he had appointed Master of the Sacred and Apostolic Palace in July 1542. Spina "planned to condemn" it. But he fell ill and died. His plan was carried out by his lifelong friend, Giovanni Maria Tolosani (c. 1471-1549). Also a friar of the Dominican Order, Tolosani wrote a treatise **On the Very Pure Truth of Divine Scripture.** He dedicated it to Pope Paul III, who ordered Spina to examine it. He praised it highly in a letter to Tolosani on 16 August 1546, shortly before he died.*

*Tolosani did what illness and death had prevented Spina from doing. In June 1544 Tolosani had completed a treatise **On the Truth of Holy Scripture**. Later he added several appendices, of which the fourth may be called, for brevity's sake, **Heaven and the Elements.** In Chapter 2 Tolosani took up the cudgels against Copernicus' **Revolutions.***

The book by Nicholas Copernicus of Toruń was printed not long ago and published in recent days. In it he tries to revive the teaching of certain Pythagoreans concerning the earth's motion, a teaching which had died out in times long past. Nobody accepts it now except Copernicus. In my judgment, he does not regard that belief to be true. On the contrary, in this book of his he wanted to show others the keenness of his mind rather than expound the truth of the matter.

Since Tolosani, other anti-Copernicans have similarly made the baseless charge that Copernicus did not believe what he wrote.

As far as I could judge by reading his book, he is a man with a keen mind. He understands Latin and Greek, and expresses himself eloquently in those languages, not however without an obscurity in his phraseology since he uses unfamiliar words too often. He is also an expert in mathematics and astronomy, but he is very deficient in physics and dialectics. Moreover he seems to be unfamiliar with Holy Scripture since he contradicts some of its principles, not without the risk to himself and to the readers of his book of straying from the faith....

Thoroughly familiar with the Bible, Copernicus farsightedly warned against allowing distortions of Biblical passages to impede the development of science.

Hence, since Copernicus does not understand physics and dialectics, it is not surprising if he is mistaken in this opinion and accepts the false as true, through ignorance of those sciences. Summon men educated in all the sciences, and let them read Copernicus, Book I, on the moving earth and the motionless starry heaven. Surely they will find that his arguments have no solidity and can be very easily refuted. For it is stupid to contradict a belief accepted by everyone over a very long time for extremely strong reasons, unless the naysayer uses more powerful and incontrovertible proofs, and completely rebuts the opposed reasoning. Copernicus does not do this at all. For he does not undermine the proofs, establishing necessary conclusions, advanced by Aristotle the philosopher and Ptolemy the astronomer.

Then let experts read Aristotle, *On the Heavens*, Book II, and the commentaries of those who have written about it...and they will find that Aristotle absolutely destroyed the arguments of the Pythagoreans. Yet this is not adduced by Copernicus in his ignorance of it, nor does he follow the Pythagoreans in all respects, since they put Fire in the middle near the center of the universe, where everybody else correctly and most convincingly proves that the earth is. Copernicus, however, puts the sun there, not Fire, and both are caught in a great error. For Copernicus puts the indestructible sun in a place subject to destruction. And since Fire naturally tends upward, it cannot, except through constraint, remain down near the center as its natural place, as the Pythagoreans falsely hold...

This difference between Copernicus and the Pythagoreans was ignored by many of Tolosani's contemporaries and coreligionists. The Pythagoreans' sun was a planet moving around the center of the universe, whereas Copernicus' sun was not a planet, did not move, and remained fixed at (or near) the center of the universe. At first, in his **Commentariolus** *Copernicus conspicuously dissociated himself from the Pythagoreans. Later on, in the* **Revolutions** *he deemed it expedient to quote or cite ancient references to the Pythagoreans. These references induced less attentive students than Tolosani to lump Copernicus indiscriminately with the Pythagoreans.*

Hence Copernicus, copying the Pythagoreans in part, leans on a cane of fragile reed which easily pierces his hand, or on an imaginary fabrication by which the truth cannot be proved. Therefore he is often mistaken. For in his imagination he changes the order of God's creatures in his system when, like the giant trying to pile Ossa on Pelion, he [seeks] to raise the earth, heavier than the other elements, from its lower place to the sphere where everybody by common consent correctly locates the sun's sphere, and to cast that sphere of the sun down to the place of the earth, contravening the rational order and Holy Writ, which declares that heaven is up, while the earth is down....

Moreover Copernicus assumes certain hypotheses which he does not prove...when he says in Book I, Chapter 8: "If anyone believes that the earth rotates, surely he will hold that its motion is natural, not violent." Copernicus assumes what he should previously have proved, namely, that the earth rotates. This proposition, however, is explicitly shown to be false. For as far as a rotating earth is concerned, its motion cannot be called natural, but [must be called] coerced, since a simple body cannot have two natural motions opposed to each other. For we see the earth move naturally toward the center [of the universe] on account of its natural heaviness. But if it is said to rotate, its circular motion will be coerced, not natural. Therefore, it is false that the earth rotates with a natural motion. On the contrary, that motion is coerced and thus Copernicus' hypothesis is completely overthrown.

Furthermore, in Book I, Chapter 10, this author falsely supposes that the "first and the highest of all [the spheres] is the sphere of the fixed stars, which contains itself and everything, and is therefore immovable." This is shown to be false, since the sphere of the fixed stars has two opposite motions, one natural, the other coerced. This could not be the case unless above it is the First Movable, which moves with a single, simple, uniform motion, as all informed astronomers agree. By the action of the First Movable, the starry heaven is moved contrary to its natural and proper motion. Copernicus would have spoken correctly, had he agreed with the theologians that above the First Movable the highest sphere is immovable, the sphere called by the theologians the Empyrean Heaven. This contains, as in an immovable place containing itself, all the lower movable heavenly spheres which revolve around the center of the universe....

Almost all the hypotheses of this author Copernicus contain something false, and very many absurditites follow from them. Hence that writer, whose name is not indicated there, and who speaks "To the Reader concerning the Hypotheses of This Work," although in the earlier part he flatters Copernicus, nevertheless toward the end of his remarks, viewing the truth of the matter correctly and without any adulation, says: "As far as hypotheses are concerned, let no one expect anything certain from astronomy, which cannot furnish it, lest he accept as the truth ideas conceived for another purpose, and depart from this study a bigger fool than when he entered it." This is what that unknown author says. These words of that author censure the book's lack of sense. For by a foolish effort it tries to revive the contrived Pythagorean belief, long since deservedly buried, since it explicitly contradicts human reason and opposed Holy Writ. Pythagoreanism could easily give rise to quarrels between Catholic expounders of Holy Writ and those persons who might wish to adhere with stubborn mind to this false belief. I have written this little work for the purpose of avoiding this scandal.

After Chapter 2, with which Tolosani originally intended to conclude his ***Heaven and the Elements,*** *he began Chapter 3 as follows:*
Although I have ended my remarks, nevertheless having been urged on by the advice of learned men, I think that some statements must still be added. For I have sent my reader to peruse the text of Aristotle, *On the Heavens,* Book II. It is not easy, however, for everybody to have that book in his own possession, together with the commentaries on it. Therefore, in order that readers may more readily learn that Nicholas Copernicus neither read nor understood the arguments of Aristotle the philosopher and Ptolemy the astronomer, for that reason I shall briefly adduce here their arguments and refutations of the opinion opposed by them....
Toward the end of Chapter 4 Tolosani remarked:

> Read Book I of Nicholas Copernicus' *Revolutions* and from what I have written here you will clearly recognize into how many and how great errors he has tumbled, even contrary to Holy Writ. Where he wished to show off the keenness of his mind in the book he published, by his own words and writings he rather revealed his own ignorance.[61]

Spina's illness and death saved Copernicus' ***Revolutions*** *from the fate that later befell Galileo's* ***Dialog concerning the Two Chief World Systems, the Ptolemaic and the Copernican.*** *Friar Tommaso Caccini, a Dominican of St. Mark in Florence, went to Rome to denounce Galileo's Copernicanism to the Inquisition. The secret testimony he gave on Friday, 20 March 1615, was later impugned. Nevertheless, he was largely responsible for the proceedings that culminated in Galileo's imprisonment and compulsory public renunciation of Copernicanism. By thus contributing to the condemnation of Galileo, Caccini helped to fulfill what had been left undone by Spina, whose friend Tolosani is an unimpeachable witness that Copernicus'* ***Revolutions*** *was not approved by the pope. Two years after the trial of Galileo, Tolosani's manuscript "was read aloud in public by Friar Tommaso Caccini, theologian of the famous Florentine Academy, in 1635."*

Reading No. 57

Osiander's First Letter to Rheticus

Andreas Osiander(1498-1552), a militant preacher of the Lutheran creed in Nuremberg, also dipped into the mathematical disciplines as his hobby. For this reason, while Rheticus was mastering Copernicus' astronomy in Frombork, he sent a letter about his teacher to Osiander. Rheticus' letter has not been preserved. But the closing part of Osiander's reply on 13 March 1540 has survived in a copy by an unknown hand. In this fragment, which was recently discovered and published for the first time, Osiander speculates about the stars and the sun - speculations which he himself characterizes as trifles (nugas). *Then Osiander continues as follows:*

...But now this is enough about these topics. What remains is that I ask you over and over again, just as you offer me your friendship, in the same way to exert your efforts so as to obtain the friendship of this man [Copernicus] for me too. For the time being, I have not dared to write to him. I also did not even want to, being certain that you would not conceal these trifles of mine from him. I [? heartily honor] him for his intellectual talents and his way of life. I also congratulate myself that up to the present time I have refrained from publishing my own material, and have not deprived him of the glory that he deserves. Farewell. Think kindly of this confused and disorderly letter of a very busy man.
Nuremberg, 13 March 1540

A Osiander[62]

Reading No. 58

Osiander Acknowledges Receiving Rheticus' *First Report*

The friends to whom Rheticus sent complimentary copies of his First Report (Gdańsk, 1540) included Osiander. His letter of acknowledgment is preserved in a fragmentary form. Only the opening lines survive, copied by an unknown scribe. The date of Osiander's acknowledgment is unclear, since it was placed at the end of the letter, which is lost. The beginning reads as follows.

To the very learned man,
Professor [George] Joachim Rheticus,
my very dear friend
Greetings, most excellent man, and very dear friend. I have

received several copies of your astronomical *[First] Report*. They have pleased me very much. The book has very clear introductory discussions of topics to be expected hereafter, such as...[63]

At this point the only surviving copy of Osiander's letter breaks off. What is preserved shows that Osiander understood that Rheticus' ***First Report*** *would be followed by a fuller exposition. Osiander looked forward to particular discussions, which he may have enumerated in the portion of the letter that has been lost.*

Reading No. 59

Osiander Challenges Copernicus' View of Astronomical Hypotheses

When Osiander was first told about Copernicus' achievement in astronomy, he did not dare to write to him. However, after receiving copies of Rheticus' ***First Report,*** *Osiander challenged Copernicus' conception of what he was doing. This letter from Osiander to Copernicus has been lost in its entirety. So has Copernicus' answer. For reasons that have not yet been clarified, Copernicus' answer, dated 1 July 1540, did not reach Osiander until March 1541. On 20 April 1541 he replied to Copernicus. What follows is the only part of his reply that has been preserved.*

I have always felt about hypotheses that they are not articles of faith but the basis of computation. Thus, even if they are false, it does not matter, provided that they reproduce exactly the phenomena of the motions. For if we follow Ptolemy's hypotheses, who will inform us whether the sun's nonuniform motion occurs on account of an epicycle or on account of the eccentricity? For, either arrangement can explain the phenomena. It would therefore appear to be desirable for you to touch upon this matter somewhat in an introduction. For in this way you would mollify the peripatetics and theologians, whose opposition you fear.[64]

Osiander's letter of 20 April 1541 to Copernicus tells us something about Copernicus' letter of 1 July 1540 to Osiander. Even though Copernicus' letter is lost, in it the astronomer must have indicated that he was afraid of trouble from the followers of Aristotle and from the theologians. The Aristotelian peripatetics still maintained that the earth is stationary. They were contradicted by Copernicus' insistence that the earth moves. The theologians taught that only those of us who are good will go to heaven after we die. But according to Copernicus, the earth is a planet revolving in the heavens around the sun. Everybody, whether good or bad, is already in heaven at birth. Aristotelianism, the dominant philosophy in the schools, was threat-

ened by Copernicus' new astronomy. Theology, whether Catholic or Protestant, was undermined by Copernicus' proclamation of the planetary nature of the earth. The traditional distinction between heaven and earth was thereby dissolved. No wonder Copernicus feared the opposition of the peripatetics and theologians. No wonder he refused for decades to publish the **Revolutions.** *No wonder he dedicated it to the pope, and published a covering letter from a cardinal.*

Reading No. 60

Osiander Suggests an Editorial Strategy to Rheticus

Osiander sent a letter to Rheticus as well as to Copernicus on 20 April 1541. The courier leaving Nuremberg carried these two letters, which complement each other. Their fate is also similar: only a part of each is preserved. What follows is the surviving part of Osiander's letter to Rheticus of 20 April 1541.

The peripatetics and theologians will be readily placated if they hear that there can be different hypotheses for the same apparent motion; that the present hypotheses are brought forward, not because they are in reality true, but because they regulate the computation of the apparent and combined motion as conveniently as may be; that it is possible for someone else to devise different hypotheses; that one man may conceive a suitable system, and another a more suitable, while both systems produce the same phenomena of motion; that each and every man is at liberty to devise more convenient hypotheses; and that if he succeeds, he is to be congratulated. In this way they will be diverted from stern defense and attracted by the charm of inquiry; first their antagonism will disappear, then they will seek the truth in vain by their own devices, and go over to the opinion of the author.[65]

Reading No. 61

The "Address to the Reader" of the *Revolutions* Contradicts what he Reads in the *Revolutions*

Before the **Revolutions** *was printed, Osiander tried to persuade Copernicus to adopt the fictionalist philosophy. Do not claim that what you are saying is true, Osiander advised Copernicus. Limit yourself to the assertion that your propositions are, or may be, useful* *for the purposes of computation. Copernicus refused to heed Osiander's well-intentioned counsel. After Rheticus' departure from Nuremberg, Osiander became the second and last editor of the* ***Revolutions****. Seizing this opportunity, among the handwritten sheets delivered to the printer presumably as fair copies of what Copernicus had composed, Osiander interpolated an "Address to the Reader," authored by himself. He deliberately kept his name off the "Address."*

To the Reader
Concerning the Hypotheses of this Work

There have already been widespread reports about the novel hypotheses of this work, which declares that the earth moves whereas the sun is at rest in the center of the universe. Hence certain scholars, I have no doubt, are deeply offended and believe that the liberal arts, which were established long ago on a sound basis, should not be thrown into confusion. But if these men are willing to examine the matter closely, they will find that the author of this work has done nothing blameworthy. For it is the duty of an astronomer to compose the history of the celestial motions through careful and expert study. Then he must conceive and devise the causes of these motions or hypotheses about them. Since he cannot in any way attain to the true causes, he will adopt whatever suppositions enable the motions to be computed correctly from the principles of geometry for the future as well as for the past. The present author has performed both these duties excellently. For these hypotheses need not be true nor even probable. On the contrary, if they provide a calculus consistent with the observations, that alone is enough. Perhaps there is someone who is so ignorant of geometry and optics that he regards the epicycle of Venus as probable, or thinks that it is the reason why Venus sometimes precedes and sometimes follows the sun by forty degrees and even more. Is there anyone who is not aware that from this assumption it necessarily follows that the diameter of the planet at perigee should appear more than four times, and the body of the planet more than sixteen times, as great as at apogee? Yet this variation is refuted by the experience of every age. In this science there are some other no less important absurdities, which need not be set forth at the moment. For this art, it is quite clear, is completely and absolutely ignorant of the causes of the apparent nonuniform motions. And if any causes are devised by the imagination, as indeed very many are, they are not put forward to convince anyone that they are true, but merely to provide a reliable basis for computation. However, since different hypotheses are sometimes offered for one and the same motion (for example, eccentricity and an epicycle for the sun's motion), the astronomer will take as his first choice that hypothesis which is the easiest to grasp. The philosopher will perhaps rather seek the semblance of the truth. But neither of them will understand or state anything certain, unless it has been divinely revealed to him.

Therefore alongside the ancient hypotheses, which are no more probable, let us permit these new hypotheses also to become known, especially since they are admirable as well as simple and bring with them a huge treasure of very skillful observations. So far as hypotheses are concerned, let no one expect anything certain from astronomy, which cannot furnish it, lest he accept as the truth ideas conceived for another purpose, and depart from this study a greater fool than when he entered it. Farewell.[66]

Reading No. 62

The City Council of Nuremberg Transmits a Less Abrasive Revision of Petreius' Answer to Tiedemann Giese's Protest against the Fraudulent Interpolation

The interpolated "Address to the Reader" enraged Rheticus when copies of the **Revolutions** *reached him in Leipzig. He sent an indignant letter, together with two copies of the book, to Copernicus' best friend, Tiedemann Giese. Whereas Rheticus blamed Petreius' wickedness, Giese suspected foul play by an underling.* ***(See Reading No. 34.)*** *Writing a letter of protest addressed to the City Council of Nuremberg, Giese asked Rheticus to transmit it. The City Council in turn forwarded it to Petreius. His answer was quite abrasive. Before sending it on to Giese, on Wednesday, 29 August 1543, the City Council instructed its secretary, Jerome Baumgartner, to tone it down.*

Wednesday, 29 August [1543]
Send to Tiedemann [Giese], bishop of Chełmno in Prussia, the answer written by Johannes Petreius to the bishop's communication (in the answer, the acerbities should be omitted and mitigated). Add to it: no punishment can be inflicted on Petreius in this matter on the basis of his answer.

Jerome Baumgartner,
Secretary to the
Council[67]

Reading No. 63

Osiander Interpolated the "Address to the Reader" without Informing Petreius

Osiander "openly admitted...that he had added this [Address to the Reader] as his own idea." He made this admission to Philip Apian (1531-1589). Apian was the son of the professor of mathematics at the University of Ingolstadt. Osiander had matriculated at that university on 9 July 1515. Long afterwards, when he was obliged to give up his post in Nuremberg for religious reasons about 18 November 1548, he may have gone back to Ingolstadt looking for another position. He may have consulted the renowned professor of mathematics there, and also conversed with his son, Philip Apian. Toward the end of January 1549, Osiander arrived in Koenigsberg. Having been appointed professor of theology at the local university, he lived in Koenigsberg until his death on 17 October 1552. Hence, the conversation between Osiander and Apian may have taken place in Ingolstadt in November or December 1548.

On 1 March 1570 Apian became professor of mathematics at the University of Tübingen. A student there purchased a copy of the first edition of the **Revolutions** *on 6 July 1570. This copy is now in the municipal library of Schaffhausen, Switzerland. Its purchaser, Michael Maestlin (1550-1631), annotated it heavily. At the top of folio 2 recto, he wrote a long note that throws much-needed light on what happened in Petreius' workshop after Rheticus left to go to Leipzig.*

At the University of Tübingen, Maestlin was Apian's assistant for a time. When Apian was dismissed for refusing to sign the oath of religious allegiance, on 23 May 1584 Maestlin replaced him. After Apian's death on 15 November 1589, Maestlin bought his predecessor's library from the widow. Rummaging through that library, Maestlin came across a historically important passage. This had been copied by Apian from an (unidentified) source. For the purpose of preserving this passage, Apian copied it into one of his own books. He did not append his signature. But Maestlin readily recognized the handwriting as Apian's. For his part, Maestlin recopied the passage in his personal copy of the **Revolutions.**

According to Apian's unknown source, Rheticus suspected that the concealed author of the Address to the Reader was Osiander, but he was not absolutely certain. To this Maestlin added: "Nevertheless, Apian told me that Osiander openly admitted to him that he had added it as his own idea."

Apian's unidentified source survives in the regional library in Munich. It is a copy of the first edition of the **Revolutions,** *which was bought for a member of the wealthy banking family in Augsburg, John Jacob Fugger (1516-1575), an ardent patron of the arts and sciences. Whoever this source was, he knew that Petreius defended* *his conduct by asserting that the Address to the Reader had been "submitted to him with the rest of the treatise" by Copernicus.*

With reference to this Address [to the Reader], I (Mich. Maestlin) found the following words written somewhere among Philip Apian's books (which I bought from his widow). Although the writer did not append his name, nevertheless I was very easily able to recognize from the formation of the letters that the handwriting was Philip Apian's. Hence, I conjecture that the following words had been copied by him from some source, doubtless for the purpose of preserving them:

> "On account of this Address [to the Reader] George Joachim Rheticus, the Leipzig professor and disciple of Copernicus, became embroiled in a very bitter wrangle with the printer [Petreius]. The latter asserted that the Address to the Reader had been submitted to him with the rest of the treatise [written by Copernicus]. Rheticus, however, suspected that Osiander had put it in the front matter of the work. If Rheticus knew this to be a fact, he declared, he would so maul the fellow that he would mind his own business, and not dare to mutilate astronomers any more in the future."

Nevertheless, Apian told me that Osiander had openly admitted to him that he had added this [Address to the Reader] as his own idea.[68]

Reading No. 64

Osiander Misled Peter Ramus

Although Osiander admitted privately to Apian that he had written the "Address to the Reader" in Copernicus' ***Revolutions,*** *he never said so publicly. The anonymity of the "Address," as printed, puzzled many readers of the* ***Revolutions.*** *Thus, Peter Ramus (Pierre de la Ramée, 1515-1572) attributed the "Address" to Rheticus. Ramus wanted to make the mathematical sciences simpler. In Rheticus'* ***Canon of the Doctrine of Triangles*** *(Leipzig, 1551), he recognized a reduction of the labor involved in trigonometrical calculations. Hence, he wrote Rheticus a letter urging him to do the same for astronomy by eliminating all its hypotheses.*

...You would accomplish this result, in my judgment, if you removed all hypotheses, and made astronomy as simple as nature itself made the essence of the stars simple. But without hypotheses (somebody may reply), the dignity of the heavenly motions cannot be retained, nor can the computation of these motions be extended.

These are, in my opinion, the two defenses of hypotheses. One of them comes from Proclus. It is inconsistent with divine bodies and unworthy of them to find the motion of the stars irregular, undefined, anomalous, so that even the collapse and disintegration of the heavens are to be feared as a result. Therefore, the greatest astronomers discovered hypotheses which would derive every revolution from reasonable, orderly, regular causes, defined by consistent calculations, and which would defend and safeguard the perpetual constancy of the heavens.

But (I say) that this nonuniformity which is under indictment is the highest uniformity, first of all in the periods and total revolutions, then in the curves and arcs, finally in individual points, and the times of approach, withdrawal, stations, retrogressions, elevations, have been individually determined and calculated in a corresponding manner in detail. Thus, at the end of Book VII of his *Laws* and in his *Timaeus,* Plato defends this decision by asserting that all the stars move around the same poles as all these motions that are observed. They do so without any irregularity (as it would appear to the authors of the hypotheses), but with marvelous regularity, which falls not a whit short of the machines in the hypotheses. Therefore, the alleged anomaly in the heavenly motions is not subject to the defect of unsteadiness. On the contrary, it displays the steadiness of the most splendid orderliness. Hence, there was no need to fear the collapse of the heavens.

But what about the question whether the computation of time cannot be extended in the absence of hypotheses? For this is (they assert) the defense of hypotheses. Here I should like you to enter into this discussion. From your resources, bring forth the materials to persuade us by a necessary inference (of a proposition which I would most gladly believe anyway) that astronomy can exist perfectly well when based exclusively on the elements and principles of arithmetic and geometry without any hypotheses.

In the first place, does it seem to you that no significance can be attached to the argument that all the motions can be observed and recorded with geometrical instruments, indeed, that they have always been so observed and recorded? But (and this is the nub of the question) has the very extension of future motions ever been inferred from some relation to past motions, without any hypotheses? Let there be a decision how [this may be done] on the basis of a history of the periods, which you know very well.

But to call upon your memory here too, of the four schools of astronomers documented by Pliny, the Babylonians, Egyptians, Greeks, Latins, I ask you to recall whether there is mention of any hypotheses of the Babylonians and ancient Egyptians and of the Greeks before Plato. Certainly, as Proclus pointed out in his *[Commentary on the] Timaeus*, Plato employed no hypotheses in astronomy. Nevertheless, when he denied that there is any disorder or confusion in the motions of the stars, he provided tne opportunity (according to the commentators on Aristotle's work *On the Heavens)* to mathematicians to investigate hypotheses by which to defend the planetary phenomena. In this way Eudoxus of Cnidus was the first to find the hypotheses of the revolving spheres. His hypotheses were corrected and improved by Aristotle, together with Callippus. Of course, in the question concerning the number of the celestial spheres, Aristotle wavered in the *Metaphysics,* Book XII. Having recourse to the astronomers as qualified judges, he mentioned no hypotheses other than those of Eudoxus and Callippus and his own, employing concentric spheres. He did not achieve satisfaction, as was clear there and also in the *Problems* dealing with the difference in latitude.

In Ramus' time, readers of Aristotle still believed that he wrote the **Problems** *in its present form. Aristotle did write a work on "Problems." Parts of it were quoted by later writers in antiquity, and are incorporated in the* **Problems** *available to Ramus and to us. But this work, as we have it, is essentially a product of the later Aristotelian School. Book XV, 4, says:*

> *The earth is the center, since the phenomena are always the same for us. This does not seem to be so, if we do not observe from the center...The earth being spherical, its center will be the same as the universe's. But we live on top of the earth. Hence, the phenomena appear to us not at the center, but half a diameter away. Then what stops the image of the phenomena from persisting when the distance increases?*

Ramus judged the **Problems'** *treatment of this question unsatisfactory. It invoked what is known as the parallactic effect. An object viewed from two nearby standpoints will be placed differently on a remote background. The parallactic effect on the moon, as observed from two widely separated stations on the earth, was known in antiquity. Copernicus' orbital motion of the earth separated out what he called the "motion in commutation" of the planets, or their parallax. Lunar and planetary parallax is detectible with the naked eye. Stellar parallax is not, requiring advanced telescopic apparatus. Hence, Copernicus' pre-telescopic* **Commentariolus,** *4th Postulate, said:*

> *The ratio of the earth's distance from the sun to the height of the firmament is so much smaller than the ratio of the earth's radius to its distance from the sun that the distance between the earth and the sun is imperceptible in comparison with the height of the firmament.*

Shortly afterwards, by order of Aristotle, Babylonian observations for 1903 years are said to have been sent to Greece by Callisthenes.

> *This misinformation was repeated by Ramus from the Venice 1540 edition of Simplicius'* ***Commentary on Aristotle's Heaven****, as translated from Greek into Latin by William of Moerbeke in Viterbo in 1271. Ramus did not consult Simplicius' Greek text. For, this said that the Babylonian observations covered 31,000 years (1000 and 10,000 × 3). For the Greek 10,000* ***(myriadon)*** *William of Moerbeke substituted 900* ***(nongentorum)****. That is why Ramus emerged with 1903. Simplicius based his statement on Porphyry (234-c. 305). Since it does not appear in any of Porphyry's surviving works, it must have been included in one of his lost works. He concocted it about six centuries after Aristotle's nephew, Callisthenes, accompanied Alexander the Great in his invasion of Asia. Babylonian observations covering 1903, not to mention 31,000, years were never cited by Aristotle anywhere in his works.*[69]

At that time there were no hypotheses. Then, according to Proclus, after the concentrics had been discarded, the Pythagoreans proposed the epicycles and eccentrics. But consider whether these Pythagoreans followed [Julius] Caesar and Sosigenes [who helped Caesar to design the Julian calendar]. Compare the origin and, so to speak, the founding of hypotheses with astronomy's age, so advanced and so long drawn out. Decide whether astronomy was formerly without hypotheses, and what convenient method can predict for a hundred and a thousand years every conjunction and effect of the stars after their motions have been observed and recorded.

For, to add what I think is most important in this regard, it seems not only completely contrary to the rules of logic but altogether impious to mingle fictions with the sacred doctrine of the heavens, especially false and absurd fictions. But the hypotheses of epicycles and eccentrics are false and absurd fictions, as is clearly shown by Venus' epicycle in your Address, if I am not mistaken, prefixed to Copernicus' *[Revolutions]*. Indeed, Proclus himself at the end of his *Hypotyposis* asserts that these hypotheses, although the most convenient of all those that had been proposed theretofore, were nevertheless imagined as a thoughtless fiction. With regard to the epicycles and eccentrics, the astronomers are seriously mistaken. They may think that the epicycles and eccentrics are only fictions because the astronomers disclose the causes of natural motions by these devices, which do not exist in the nature of things. Or they may suppose that reality is faulty. For they shatter the connected series of the celestial globes by assigning one motion to the epicycles and eccentrics, and another to their planet, and by dividing, combining, and separating the bodies in various ways. Furthermore, Proclus says that the causes of the planes and intervals are not assigned in any way by the astronomers who resort to hypotheses. Hence, Proclus concludes, by putting the cart before the horse such astronomers do not derive their conclusions from the hypotheses, as they should have done following the example of the other sciences, but they derive the hypotheses from the conclusions.

Therefore, conceive of so important an element of so great a science as a Gordian knot to be untied or at least cut. As the reward for the solution, believe that what has been proposed is the domination, not over Asia, but over astronomy.

Furthermore, do not be afraid that somebody may say the removal of hypotheses will cause the destruction of a goodly supply of very beautiful proofs. For, beauty produced by false colors adds no adornment to nature's true mien and charm. On the contrary, it absolutely deforms and corrupts them.

Therefore, by those gods whose homes and temples we are discussing, I beseech you, undertake the task most deserving of your outstanding skill to permit astronomy, freed by you from fictional hypotheses, to display to men's minds the lights of its stars just as splendid as nature presented them to be contemplated by our eyes. There is no greater service than this, in my judgment, by which you may bind mankind to the perpetual celebration of your name.

Farewell. If you answer me, as I fondly hope, send your letter to [Joachim] Camerarius, a very learned man, most friendly to you, so that it may be forwarded more securely by him to me. Again, farewell.

Paris, Collège de Presles,
25 August 1563

Most devoted to you,
your Peter Ramus,
Royal Lecturer[70]

Reading No. 65

Ramus Offers to Resign from his Royal Lectureship to Make Way for Any Astronomer who Would Discard the Hypotheses

After appealing privately to Rheticus in 1563, in his Scholae mathematicae (Paris/Basel, 1567-1569) Ramus issued a general call to the German astronomers to establish an astronomy without hypotheses. He promised a royal lectureship in Paris to whoever succeeded. If necessary, he would resign from his own royal lectureship. That did become vacant, but not by voluntary resignation. For on 24 August

1572 Ramus was slaughtered, together with many of his fellow Protestants, in the horrible massacre of St. Bartholomew's Day.

...The fiction of hypotheses is therefore absurd. Yet it is more naive in Eudoxus, Aristotle, and Callippus, who thought their hypotheses were true. Indeed, they venerated them as though they were the gods of the starless spheres. But in later astronomers the fable is far and away the most absurd: to demonstrate the truth of natural events by false causes. Therefore, logic in the first place, then mathematics, the elements of arithmetic and geometry, will contribute the greatest help in establishing the purity and dignity of the most inclusive science. Would that Copernicus had preferably stumbled on that thought of constituting astronomy without hypotheses! For it would have been far easier for him to delineate an astronomy corresponding to the truth of his stars than, as with some mighty effort, to make the earth movable, so that we might watch the motionless stars while the earth moves. Nay rather from so many magnificent universities of Germany may a philosopher who is likewise an astronomer arise to win the centrally placed prize of eternal glory. If a fruit of fleeting usefulness can be proposed as the reward for such great excellence, I promise you a royal lectureship in Paris as the prize for an astronomy fashioned without hypotheses. For my part, I shall fulfill this promise most gladly, even by resigning from my lectureship.[71]

Reading No. 66

Kepler Answered Ramus' Call, while Disclosing who Wrote the "Address to the Reader" in Copernicus' *Revolutions*

Ramus told the academic world how Rheticus responded to his private letter of 25 August 1563:

> *Rheticus...made Leipzig and also Cracow famous for mathematics. As a result of my letter looking to the liberation of astronomy from hypotheses, he would have provided hope of making the University of Paris famous too. Had he not been compelled to master medicine and practice it in the guise of a patron, mathematics would long since have been celebrating a second Copernicus.*[72]

No known answer came from the German astronomers to Ramus' public call in 1567 and 1569. Soon thereafter his enemies put an end to his life and his lectureship. When Ramus was assassinated, Kepler was less than a year old. Nevertheless, Ramus' call was reprinted on the verso of the title page of Kepler's New Astronomy (Heidelberg, 1609). In replying to Ramus there, Kepler dissociated Copernicus from the "Address to the Reader" in the ***Revolutions.*** *No fictionalist, Copernicus believed that what he said was true. In that case, who was responsible for the "Address to the Reader"? Kepler made the first public disclosure that Osiander was the guilty party. Kepler's quotation from Ramus ended with his promise to resign from his royal lectureship in favor of the successful astronomer.*

You abandoned this pledge, O Ramus, just in time by giving up your life and your lectureship. If you held this lectureship now, I would rightfully claim it for myself. By virtue of this work [Kepler's *New Astronomy]* I shall prevail, even when your logic is the judge. You demand help for the most extensive science [astronomy] only from logic and mathematics. Please do not exclude the help of physics, which it cannot do without. Unless I am mistaken, you surrender easily, since you enfold your system-builder in philosophy as well as in mathematics. Accordingly with equal readiness heed philosophy defending something that is most absurd by common repute, not with a gigantic effort, but with very good reasons. When it does so, it does nothing new, nothing unfamiliar, but it fulfills its function, for which it was devised.

It is a most absurd fable, I admit, to demonstrate natural phenomena by means of false causes. Yet this fable is not in Copernicus. For he himself thought that his hypotheses were true, no less than your ancient writers thought theirs to be true. Not only did he think so, but he also proves that they are true. As witness, I offer this work.

But do you wish to know the contriver of this fable, which angers you so much? Andreas Osiander is noted in my copy in the handwriting of Jerome Schreiber of Nuremberg. Therefore, that "Address to the Reader," which you say is most absurd, was deemed very prudent by this Andreas, when he was supervising the edition of Copernicus (as can be inferred from his letter to Copernicus) and he put it in the front matter of the book while Copernicus himself was either already dead or at least unaware [of what was going on in Nuremberg]. Therefore, Copernicus is not creating a myth, but is dissenting, that is, he is philosophizing, what you were asking for in an astronomer.[73]

Kepler did not know whether Copernicus was already dead when Osiander was interpolating the "Address to the Reader." We are now certain that Copernicus was still alive then. For, a copy of the ***Revolutions*** *was sent from Nuremberg, where it was printed, to Emperor Charles V on 21 March 1543. Copernicus died a little more than two months later, on 24 May 1543.*

When did Osiander interpolate the Address? Rheticus, the first editor of the ***Revolutions,*** *left Nuremberg early in October 1542. At that time the typesetting of Books I-IV had been completed. Books V-VI and the front matter remained. In the absence of precise infor-*

mation, a guess may be hazarded that Osiander composed and interpolated the Address early in March 1543. Copernicus was then a very sick old man in distant Frombork. He was, as Kepler correctly surmised, unaware of what Osiander was doing.

Kepler tells us how he learned that Osiander was the author of the interpolated Address. This information was obtained by Kepler from the copy of the **Revolutions** *that was presented by the publisher Petreius to Jerome Schreiber of Nuremberg. Schreiber had entered the University of Wittenberg in the summer semester of 1531 as a classmate of Rheticus. When Rheticus went on leave in May 1542, he was replaced as professor of astronomy in the University of Wittenberg by Schreiber. On 1 October 1543 Schreiber left Wittenberg to travel in Italy. On the way south, he visited his mother in Nuremberg, and called on old friends in his birthplace. These included Petreius, who presented Schreiber with a copy of the* **Revolutions.** *At that time Schreiber may have been told about Osiander's authorship of the Address by Petreius. Or he may have been told by Jerome Baumgartner, who was the Secretary of the City Council of Nuremberg on 29 August 1543 when it acted on Petreius' answer to Tiedemann Giese.* **(See Reading No. 62.)** *Schreiber carried a letter of recommendation, dated 1 October 1543, from Melanchthon to Baumgartner.*

When Schreiber left Nuremberg to resume his trip to Italy, he deposited his belongings in his mother's house. From Italy, Schreiber extended his journey to France. Troubled by an old ailment, he died in Paris in 1547. The library he had left behind in Nuremberg was sold to a local book dealer.[74] *Kepler, a distant relative of that dealer, had Schreiber's astronomical books and manuscripts when he went off to study that subject at the University of Tübingen. That is how Kepler was able to verify that the hand that wrote ANDREAS OSIANDER in capital letters over the Address to the Reader in the Petreius-Schreiber-Kepler copy of the* **Revolutions** *was Schreiber's. A facsimile of that copy, now in the library of the University of Leipzig, was published in 1956 by Edition Leipzig and Johnson Reprint Corporation (New York/London).*

Reading No. 67

Copernicus' Reason for Breaking Away from the Ptolemaic Tradition

Ptolemy's equant (Chapter 2, Figure 7) allowed him to bring his planetary theory into closer agreement with his observational data than had any of his predecessors. He made this advance, however, at the cost of violating the rule that motion on a circle must be uniform as viewed from the center of the circle. In Ptolemy's equant, motion on the circle was at a uniform distance from the center of the circle.

But as viewed from the center, the motion was nonuniform. It was uniform, however, as viewed from the equant point outside the center.

Ptolemy's departure from the rule of uniform motion disturbed Copernicus. He looked for an alternative way to account for the observed nonuniformity while adhering to the principle of uniform motion. This is how, according to his own account, he came to consider the earth as movable. Only thereafter did he undertake "the task of rereading the works of all the philosophers which [he] could obtain to learn whether anyone had ever proposed other motions of the universe's spheres than those expounded by the teachers of astronomy in the schools" (Copernicus, ***On the Revolutions,*** *p. 4/42-45).*

As the ancient astronomers admit, a circular motion can be uniform with respect to an extraneous center not its own....And now in the case of Mercury the same thing is permitted, and even more. But (in my opinion) I have already adequately refuted this idea in connection with the moon. These and similar situations gave me the occasion to consider the motion of the earth and other ways of preserving uniform motion and the principles of the science, as well as of making the computation of the apparent nonuniform motion more enduring.[75]

Reading No. 68

The Unique Scientific Revolution

Science today is quite different from what it used to be. The greatest scientists of the past devoted themselves most intensively to the study of nature in all its aspects as a mathematician sought to discover the unknown properties of numbers and geometrical figures. The pleasure derived by a scientist from his unraveling of a secret of nature was like his delight in viewing a splendid sunset. Throughout this period of contemplative science, the hewers of wood and the drawers of water received little guidance from the theoretical investigations of the scientists.

The crucial turn came when the scientists scrutinized the operations of the workmen. The hand was united with the brain. Science became operational in its outlook. The difference between the older and newer attitudes toward science was clearly delineated by Robert Boyle (1627-1691), who was both a skilled practitioner and a talented thinker.

There are two very distinct ends, that men may propound to themselves in studying natural philosophy [as science was then called]. For some men care only to know nature, others desire to command her; or, to express it otherwise, some there are, who desire but to please themselves by the discovery of the causes of the known phenomena; and others would be able to produce new ones, and bring nature to be serviceable to their particular ends, whether of health, or riches, or sensual delight....[The latter] have produced inventions of greater use to mankind, than were ever made by *Leucippus*, or *Epicurus*, or *Aristotle*, or *Telesius*, or *Campanella*, or perhaps any of the speculative devisers of new hypotheses; whose contemplations aiming for the most part but at the solving, not the increasing or applying, of the phenomena of nature, it is no wonder they have been more ingenious than fruitful, and have hitherto more delighted than otherwise benefitted mankind.[76]

Footnotes to Part II

1. *Commentaria in Aristotelem graeca,* VII (Berlin, 1884), ed. J.L.Heiberg, *Simplicii in Aristotelis De caelo commentaria,* p. 488/18-24, p. 492/31 - 493/5.

2. Copernicus never taught mathematics in Germany.

3. Nicholas Copernicus, *On the Revolutions,* translation and commentary by Edward Rosen (Baltimore, 1978), Book III, Chapter 15, p. 155/4 - 156/45.

4. *Ibid.,* Book I, Chapter 8, p. 16/38 - p. 17/27.

5. Leopold Prowe, *Nicolaus Coppernicus* (Berlin, 1883-1884; reprint, Osnabrück, 1967), I, 1, 267.

6. *Ibid.,* II, 516-517. A modification of this provision of the statutes in 1540 was published in *Zeitschrift für die Geschichte und Altertumskunde Ermlands,* 1972, *36,* Sonderdruck: *Die Statuten des ermländischen Domkapitels aus dem Jahre 1532 und ihre Novellierungen,* p. 112.

7. Prowe, I, 1, 291.

8. One "C" is missing in Prowe's date.

9. Prowe, I, 1, 291.

10. Erice Rigoni, "Un autografo di Niccolo Copernico," *Archivio veneto,* 1951, anno 81, vol. 48-49, 5th series, #83-84, pp. 147-150; Hans Schmauch, "Des Kopernikus Beziehungen zu Schlesien," *Archiv für schlesische Kirchengeschichte,* 1955, *12:* opposite p. 154; "Um Nikolaus Copernicus," *Studien zur Geschichte des Preussenlandes,* Festschrift für Erich Keyser, ed. Ernst Bahr (Marburg, 1963), opposite p. 425.

11. Baldassarre Boncompagni, "Intorno ad un documento inedito relativo a Niccolo Copernico," *Atti dell' Accademia Pontificia de' Nuovi Lincei,* 1877, *30:* 341-397; facsimile between pp. 398-399.

12. Prowe, I, 1, 335, 337.

13. *Ibid.,* I, 2, 212/10-20; II, 512-513, #38.

14. Canon Zacharias Tapiau in 1508, shortly before his death on 20 January 1509, bequeathed 522 1/2 marks for the establishment of two vicariates; *Zeitschrift für die Geschichte und Altertumskunde Ermlands*, 1916, *19*: 817;1931, *24*: 57/6.

15. Prowe, I, 1, 381/last 7 lines; I, 2, 256/3: 238 7/8 marks.

16. Jerzy Drewnowski, *Mikołaj Kopernik w świetle swej korespondencji* (Wrocław, 1978; Studia Copernicana, *18*), pp. 228-229, #3.

17. *Ibid.*, p. 233, #8. Copernicus dated this letter from Frombork, for which he coined the Greek equivalent Gynopolis, since the town's name in German is Frauenburg (Citadel of Our Lady). In writing to Dantiscus, a well-known neo-Latin poet and classical scholar, Copernicus felt no hesitation in identifying his place of residence as Gynopolis. He matched Frombork with Gynopolis also in *Revolutions*, III, 13, and IV, 7 (Copernicus, *On the Revolutions*, p. 145/13-14, 191/29-30).

18. Ludwik Antoni Birkenmajer, *Mikołaj Kopernik* (Cracow, 1900), pp. 392-393, #8.

19. Marian Biskup, "Sprawa Mikołaja Kopernika i Anny Schilling w świetle listów Feliksa Reicha do Biskupa Jana Dantyszka z 1539 roku," *Kommunikaty Mazursko-Warmińskie*, 1972, p. 375/14 - 376/14.

20. *Ibid.*, p. 378/1 - 379/8.

21. Jerzy Drewnowski, "Nowe źródło do niedoszłego procesu kanonicznego przeciwko Mikołajowi Kopernikowi," *Kwartalnik historii nauki i techniki*, 1978, *23:* 184/13-35; facsimile on p. 180 and in Drewnowski, Studia, Plate 8.

22. L.A.Birkenmajer, *M. Kopernik*, pp. 393-394, #10.

23. *Ibid.*, p. 394, #11; Chełmża (Kulmsee, in German), according to Marian Biskup, *Regesta Copernicana* (Studia Copernicana, VIII; Wrocław, *1973), p. 179, #415.*

24. Prowe, I, 2, p. 301, n.

25. L.A. Birkenmajer, M. Kopernik, p. 395, #13.

26. Ibid., p. 396, #14.

27. Prowe, I, 2, pp. 366-367, n.

28. Jeremi Wasiutyński, *Kopernik, twórca nowego nieba* (Warsaw, 1938), pp. 591-592, n. 200.

29. Jeremi Wasiutyński, "Uwagi o niektórych kopernikanach szwedzkich," *Studia i materiały z dziejów nauki polskiej,* Series C, 1963, VII, 77-78.

30. *Ibid.,* pp. 79-80.

31. Prowe, II, 418-419.

32. *Ibid.,* I, 2, 561.

33. *Ibid.,* II, 419-421.

34. *Ibid.,* I, 2, pp. 369-370.

35. *Ibid.,* p. 370.

36. H.D.MacLeod, *Elements of Political Economy* (London, 1858; 1st ed., 1857), pp. 475-477.

37. Raymond de Roover, *Gresham on Foreign Exchange* (Cambridge, MA, 1949), p. 91.

38. *Ibid.,* p. 93.

39. *Joannis Regiomontani Opera collectanea,* ed. Felix Schmeidler (Osnabrück, 1972), p. 60/12up - 9up; facsimile of the Venice 1496 edition of the Peurbach-Regiomontanus edition of the *Epitome,* sig. a2v.

40. Maximilian Curtze, "Der Briefwechsel Regiomontan's mit Giovanni Bianchini, Jacob von Speyer und Christian Roder," *Abhandlungen zur Geschichte der mathematischen Wissenschaften,* 1902, *12:* 327/22-26.

41. Rinaldo Fulin, "Documenti per servire alla storia della tipografia veneziana," *Archivio veneto,* 1882, *23:* 119 #40.

42. *Regiomontani Opera collectanea,* ed. Schmeidler (n. 39), p. 145/7-13.

43. Rosen, *Three Copernican Treatises,* p. 57 (slightly modified).

44. A.I.Sabra, "An Eleventh-Century Refutation of Ptolemy's Planetary Theory," pp. 121-122, n. 13, in *Science and History, Studies in Honor of Edward Rosen* (Studia Copernicana XVI; Wrocław, 1978), ed. Erna Hilfstein et al.

45. *Regiomontani Opera collectanea* (n. 39), p. 105/27-42 (*Epitome,* III, 13).

46. Copernicus, *On the Revolutions,* p. 162/7-45 (III, 20).

47. *Al-Battani sive Albatenii Opus astronomicum,* ed. Carlo Alfonso Nallino (Milan, 1899-1907; Pubblicazioni del Reale Osservatorio di Brera in Milano, #40), I, 44/17-37.

48. Rosen, *Three Copernican Treatises,* pp. 77-78 (somewhat modified).

49. Copernicus, *On the Revolutions,* p. 240/37-242/2.

50. Burmeister, *Rhetikus,* III, #7, p. 39/17-28.

51. *Ibid.,* #8, p. 40.

52. *Luther's Works,* vol. 54, *Table Talk,* ed. tr. Theodore G. Tappert (Philadelphia, 1967), pp. 358-359.

53. Copernicus, *On the Revolutions,* p. 11/45-12/11.

54. *Ibid.,* p. 16/9-20.

55. Reprinted in Prowe, II, 288/10 up - 9 up.

56. Enrico Stevenson, Jr., *Inventario dei libri stampati palatino-vaticani* (Rome, 1886-1889), I, part 2, page 161, #2250.

57. Ernst Zinner, *Entstehung und Ausbreitung der coppernicanischen Lehre* (Erlangen, 1943), p. 451.

58. Copernicus, *On the Revolutions,* p. 5/23 - 6/9.

59. *Ibid.,* p. 3.

60. *Ibid.,* p. XVII.

61. Eugenio Garin, "Alle origini della polemica anticopernicana," in *Studia Copernicana* VI = Colloquia Copernicana II (Wrocław, 1973), 35/14 up - 2 up, 36/8-14 up, 37/3-11, 37/15 - 38/21, 41/last 5 lines.

62. Martha List, "Marginalien zum Handexemplar Keplers von Copernicus: *De revolutionibus orbium coelestium* (Nürnberg, 1543)," in *Science and History,* Studia Copernicana XVI, p. 456/10-18.

63. *Ibid.,* p. 456/21-26.

64. Rosen, *Three Copernican Treatises,* pp. 22-23 (somewhat modified).

65. *Ibid.,* p. 23.

66. Copernicus, *On the Revolutions,* p. XVI.

67. L.A. Birkenmajer, *M. Kopernik,* p. 403.

68. Zinner, *Entstehung,* p. 453.

69. *Commentaria in Aristotelem graeca,* VII, ed. J.L. Heiberg (Berlin, 1894), p. 506/11-14.

70. *Petri Rami...collectaneae praefationes, epistolae, orationes* (Hildesheim, 1969; reprint of the Marburg 1599 edition), p. 215/14-218. Ramus' letter to Rheticus was printed for the first time in the preface of Ramus' posthumous *Professio regia* (Basel, 1576).

71. Ramus, *Prooemium mathematicum* (Paris, 1567) and *Scholae mathematicae* (Basel, 1569), II, 50.

72. *Ibid.,* p. 66/12 up-7 up.

73. Johannes Kepler, *Gesammelte Werke,* III (Munich, 1937), 6/19-38; Edward Rosen, "The Exposure of the Fraudulent Address to the Reader in Copernicus' *Revolutions,*" *Sixteenth Century Journal,* 1983, *14:* 283-291.

74. List (n. 62, above), p. 451.

75. Copernicus, *On the Revolutions,* V, 2; p. 240/26-33.

76. Robert Boyle, *Certain Physiological Essays,* in *The Works of the Honourable Robert Boyle* (London, 1744), I, 199.

SUGGESTIONS FOR FURTHER READING

Alberuni's India, tr. Edward C. Sachau, 2 vols. (Lahore, 1962; reprint of London 1910 ed.)

Birkenmajer, Ludwik Antoni, *Mikołaj Kopernik* (Cracow, 1900); the partial English translation (Xerox University Microfilms, 1976) should be used with extreme caution

Burkert, Walter, *Lore and Science in Ancient Pythagoreanism*, tr. from German by Edwin L. Minar, Jr. (Cambridge, MA, 1972)

Burmeister, Karl Heinz, *Georg Joachim Rhetikus*, 3 vols. (Wiesbaden, 1967-1968)

Campanus of Novara and Medieval Planetary Theory: Theorica planetarum, eds. trs. Francis S. Benjamin, Jr. and G. J. Toomer (Madison/Milwaukee/London, 1971)

Carmody, Francis J., *Arabic Astronomical and Astrological Sciences in Latin Translation: a Critical Bibliography* (Berkeley/Los Angeles, 1956)

Cohen, Morris Raphael and Israel Edward Drabkin, *A Source Book in Greek Science* (Cambridge, MA, 1948, 1966)

Copernicus, Nicholas, Complete Works, I (London/Warsaw, 1972); II (Baltimore, 1978); III (Warsaw, 1984)

Dicks, David R., *Early Greek Astronomy to Aristotle* (Ithaca, 1970)

Dictionary of Scientific Biography, 16 vols. (New York, 1970-1980)

Dreyer, John Louis Emil, *A History of Astronomy from Thales to Kepler* (New York, 1953; modified reprint of the Cambridge 1906 ed.)

Duhem, Pierre, *Le Systeme du monde*, 10 vols. (Paris, 1913-1959)

Duhem, Pierre, *To Save the Phenomena: an Essay on the Idea of Physical Theory from Plato to Galileo*, tr. from French (1908) by E. Doland and C. Maschler (Chicago, 1969)

Goldstein, Bernard R., *Ibn al-Muthanna's Commentary on the Astronomical Tables of al-Khwarizmi* (New Haven/London, 1967)

Grant, Edward, ed. *A Source Book in Medieval Science* (Cambridge, MA, 1974)

Heath, Thomas Little, *Aristarchus of Samos* (Oxford, 1959; reprint of Oxford 1913 ed.)

Hodson, F. R., ed., *The Place of Astronomy in the Ancient World* (London, 1974)

Kuhn, Thomas S. *The Structure of Scientific Revolutions,* 2nd ed. (Chicago, 1970)

Kunitsch, Paul, *Der Almagest, Die Syntaxis Mathematica des Claudius Ptolemäus in arabisch-lateinischer Ueberlieferung* (Wiesbaden, 1974)

Lloyd, G. E. R., *Aristotle: The Growth and Structure of His Thought* (Cambridge, England, 1968)

Maimonides, Moses, *The Guide of the Perplexed,* tr. Shlomo Pines (Chicago, 1963)

Needham, Joseph, *Science and Civilisation in China,* III (Cambridge, England, 1959)

Neugebauer, Otto, *A History of Ancient Mathematical Astronomy,* 3 vols. (New York, 1975)

Neugebauer, Otto, *The Exact Sciences in Antiquity,* 2nd ed. (New York, 1962)

North, John David, *Richard of Wallingford,* 3 vols. (Oxford, 1976)

Oresme, Nicole, *Le Livre du ciel et du monde,* ed. tr. Albert D. Menut and Alexander J. Denomy (Madison/Milwaukee/London, 1968)

Pannekoek, Antonie, *A History of Astronomy* (New York, 1961; tr. from Dutch ed., Amsterdam/Antwerp, 1951)

Pedersen, Olaf, *A Survey of the Almagest* (Odense, 1974)

Prowe, Leopold, *Nicolaus Coppernicus* (Osnabrück, 1967; reprint of Berlin, 1883-1884 ed.)

Rosen, Edward, *Three Copernican Treatises,* 3rd ed. (New York, 1971)

Solmsen, Friedrich, *Aristotle's System of the Physical World* (Ithaca, 1960; reprint, Johnson Reprint Corp.)

Studia Copernicana, I-XXI (Wrocław, 1970-1982)

Westman, Robert S., *The Copernican Achievement* (Berkeley/Los Angeles/London, 1975; reviewed in the *Polish Review,* 1976, *21:*225-235)

Zinner, Ernst, *Entstehung und Ausbreitung der coppernicanischen Lehre* (Erlangen, 1943)

Zinner, Ernst, *Leben und Wirken des Joh. Müller von Königsberg genannt Regiomontanus,* 2nd ed. (Osnabrück, 1968)

INDEX